TORPEDO 8

TORPEDO 8

THE STORY OF SWEDE LARSEN'S

BOMBER SQUADRON

BY

IRA WOLFERT

The P-47 Press
2019

CONTENTS

FOREWORD

The Japs wiped out the United States Navy Torpedo Squadron 8 in a few minutes at the Battle of Midway [4-7 June 1942; six months after the Pearl Harbor attack]. The minutes were hot and rough. The squadron was like a raw egg thrown into an electric fan, and only three men came out of the action alive. One of these is no longer fit for combat duty. His nerves are gone. They became unstrung in those few minutes, and in the ten months since then he has not been able to get them working again normally, although he has been out on the line trying his best, refusing painfully to give up.

Our Navy, too, has wiped out whole Jap squadrons in a few rough, hot minutes, and there has always been great curiosity on our part as to how the Jap airmen left to carry on were taking it.

When, as happened late in October over Guadalcanal, two Jap squadrons of high-level bombers took off for a routine bombing of a not especially important target and only a single airplane returned; when, a few days later, Jap torpedo bombers launched an attack on the carrier Hornet with nearly as many torpedo planes as they had used against the Pacific Fleet at Pearl Harbor December 7, and not a single airplane survived; when, two weeks later, two squadrons of Jap torpedo planes, a force as large as the one used at Pearl Harbor, were destroyed to the last plane in a few seconds more than nine minutes; when these hammer blows fell, how did the comrades of the dead feel, the men who had to step into the dead men's shoes and follow down their awful path?

What was their reaction when they heard this news and how did they feel when they were ordered into similar planes and locked themselves in to take off on similar missions? Did they have left in them any of the confidence in survival necessary for the efficient operation of complicated machinery and necessary for the pressing home of an attack?

Or did they have in them the unnerving efficiency-destroying emotions of men ordered to commit suicide and determined to commit suicide?

We wanted to know. We were most anxious to know, and on November 12, toward the end of a bloody afternoon on Guadalcanal, we found out. That was the day two squadrons of Jap torpedo bombers were obliterated.

Two young Jap airmen had managed to live unhurt through the destruction of their plane and had been taken prisoner. They were both asked how they had felt when twenty-five bombers were sent out and one came back, when twenty Zeros [Mitsubishi A6M long-range fighters] were sent out and none came back, when thirty-three planes were sent out and only one came back, when, day after day, on mission after mission, it was only the odd planes that survived.

They answered with enough difference to show that they had not been coached and that they were answering honestly, to the best of their ability. They did not blame their planes, their training, or their officers, or develop any great fear of us. Instead, each said he felt, as all Jap airmen felt, that the dead were responsible for their own deaths; they had not done their work properly

Americans do not blame the dead for their deaths. Our traditions and teachings are against it. The fascists—and the Japanese are purer fascists than the Germans or Italians, each of whom has had something of a democratic tradition— cannot seem to conceive of their state or their superiors being wrong, but automatically think of themselves as being wrong.

So, they blame their dead for having died. But we, as democrats, have too much respect for the individual man to blame, without overwhelming evidence, our dead for having died. This reaction is as automatic in us as the fascist reaction is in them. Also, we are too realistic and too sensitive to the worth of each individual man to accept a state or those in authority over us as being infallible.

Nor do we have any inclinations for suicide or for suicidal actions in war.* Our traditions, our teaching, and the whole American temperament are all against such antics.

So, when Torpedo 8 was wiped out on Thursday morning, June 4, 1942, in about the time it takes to stamp out a pile of ants, it looked to those of us on the outside as if torpedo bombing were about to become a lost art.

That, those of us on the outside felt, was too bad because no man has ever thought of a more devastating weapon against a ship at sea than a torpedo, and the airplane has numerous magnificent advantages over the submarine and the destroyer in delivering torpedoes [primarily, accuracy].

But it is against our idea of how to fight a war to buy expensive planes* for one-way rides and send men out who know they have no chance to come back. We are against it, if for no other reason than that it is inefficient. So, goodbye to torpedo plane. We on the outside consigned them to that grave for dead notions already occupied by the man-steered torpedoes the Italians had developed with such poor results. But the Navy did not agree. Nor did Torpedo 8 agree.

*America produced over 300,000 aircraft during World War 2.

The Navy seemed to know without asking that Torpedo 8 would not feel this way, for, without being asked, Torpedo 8 was thrown directly from Midway* into the Battle for the Solomons, a series of engagements into which the Japs put about five times the naval strength they used at Midway, and much more naval strength than they used against the Malay Peninsula and Java.

*The Battle of Midway came a month after the Battle of the Coral Sea, 4 to 8 May 1942, which was the first carrier-to-carrier battle in Naval history, the first naval conflict in which neither ship fired at each other and the first instance of airplane combat over the Pacific Ocean.

Torpedo 8 went into the battle with two veterans of Midway, plus remnants of the old squadron who had not got into the action there, and plus "replacements," as they are called. They did not, as the Japs do, blame their dead for having died. They wanted revenge for them. Up to Midway, the slogan of the squadron had been "Attack." On June 12, eight days after the holocaust at Midway, the squadron commander in an official squadron memorandum changed the slogan to: "ATTACK AND VENGEANCE!"

These deadly young Americans, committed like characters in romantic fiction to revenge for their dead comrades, exacted a vengeance greater than any fiction writer would dare imagine. In three months and one week, they carried out thirty-nine attack missions, sixteen against ships, twenty-three against ground targets.

They were credited with two carriers. They also hit a battleship, five heavy cruisers, four light cruisers, one destroyer, and one transport. The Japs they killed to make up for their forty-two dead at Midway cannot be counted accurately, but must run into the thousands.

They fought the enemy's ships, planes, and troops. They strafed, and did dogfighting. When there were no Jap ships to torpedo, they glide-bombed Japs on the ground. When the Japs blew up all their planes, and they had to wait for more to be sent, they took tommy-guns and rifles and helped the Marines hold the line, and when they were relieved there, they went out sniper hunting, and found snipers and killed them.

Many on Henderson Field [the contested airfield on Guadalcanal, which the Japanese called Lunga Point] knew what those fellows were up to and tried to help them keep score. It was impossible to be anything like accurate, since most of their work was carried out hundreds of miles within Jap territory. But once the Marines went hotfooting into Jap lines on Guadalcanal on the heels of a routine softening-up by Torpedo 8, an accurate count could be had, for the Marines had to bury the Jap dead.

They buried four hundred and seven Japs who had died of Torpedo 8's bombs. This was after only one of twenty-three such missions. Torpedo 8's revenge is one of the most grimly satisfying events of the American part of the war thus far. It's a legend of the kind that made epic ballads in ancient days.

And this is the story of the men who lived that legend, of why they went out for revenge, and how they got it, and how they felt while getting it. I knew some of the men on Guadalcanal and watched them in action there. I don't think I'll ever forget watching their fat little planes going out to attack, wagging their guns, or ever forget waiting for them to come back and counting them silently as they came back.

Because of a shift in nomenclature, ordained by the unromantic facts of war and of Navy arithmetic, there is no more Torpedo 8. It has been decommissioned and its men have been scattered to salt down and pepper up other torpedo squadrons.

But I saw some of the story as it was being lived on Guadalcanal during the latter part of 1942. Then during four concentrated days in Washington with Lieutenant Harold H. Larsen, who sat talking with the squadron's official War Diary on his lap, I got the rest of it, working back to its starting point early in the same year. And this, more or less, rather more than less, is how the story goes.

George Gay (right), sole survivor of VT-8's Destruction at Midway

1 - TORPEDO 8 IS COMMISSIONED

THE NEW TORPEDO PLANES took off from Norfolk, Virginia. This was in May, two months after the manufacturer had delivered them to the men of Torpedo 8, who had been detached from the aircraft carrier Hornet to put the planes into commission. Putting new type planes into commission for combat in two months is a record.

The Navy had never done it before nor had the Army, nor, so far as is known, had the army or navy of any other nation in the world. The reason for the haste was that this new plane—TBF, the Navy called it, and the manufacturer, having Pearl Harbor in mind, called it the Grumman Avenger—was such a beauty and handled so well and so fast that the men working on it hated to think of their friends back on the Hornet having to go into action without it.

Then, the fellow in charge, Lieutenant Harold H. Larsen, who, at the time, was Torpedo 8's executive officer, was a driver. He was a real hard man to work for. He admitted it.

"I'm not very smart," he said. "I've got brains that work slow and the only smart thing I ever thought of in my life is that, to realize that I'm not very smart, and if there's anything in life I want, I have to work for it, work like hell for it because I haven't the brains to get it easily, like some fellows. It's become a habit with me now to be that way, to work like hell for anything I want."

What Lieutenant Larsen wanted was to get the war over with. That was his idea, get into the war and get it over with, and do your fooling around later when it would be fun, when you would have nothing hanging over your head to make you think that maybe today's fun was not going to have any tomorrow.

So he was down in the pits or up in the air all the time, grease on him, snapping the whip, and when the fellows wanted to knock off for an hour or so to watch Bob Feller pitch and said, "Hell, Swede, the war'll keep that long," there was a look of surprise on his face that you could see plainly through the grease.

Swede is out of Omaha originally, a big blond straight-haired man with the build and the walk of an old time sergeant-major. He was thirty at the time, and he liked baseball as well as anybody could.

He had played it at the Naval Academy, along with football and lacrosse. But he was a fellow who had to think of one thing at a time. Swede had a soft voice, yet when he snapped the whip you sure heard it crack, and the fellows didn't get to see Bob Feller∗ and the Navy's "Tunneyfish," as Gene Tunney's∗∗ boys were called, mow down the opposition. They flew over the ballpark once or twice, but that's the nearest they got.

∗Baseball star Robert William Andrew Feller (1918-2010) was the first American professional athlete to enlist in the services during World War 2.

∗∗Legendary World Heavyweight Boxing Champion James Joseph "Gene" Tunney (1897-1978) was inducted into the United States Marine Corps Sports Hall of Fame in 2001.

As all the fellows admit, Swede is harder on himself than on anybody. When his Uncle Karl Stefan got elected to Congress from Nebraska, Swede gave up seeing him. He didn't want Uncle Karl to think he was hanging around expecting favors. He didn't see Uncle Karl for ten years, not until there was nothing anymore that Uncle Karl could do for him, because he had become a man with a career that only God could do anything about. And he was hard on his own football career. A football career is something that grows bigger, stronger, and more glorious in the memory of the average fellow who's had one as the years pass.

"But Swede, who got into six varsity games one season and missed winning his 'N' by a few minutes of play, describes his adventures on the gridiron in this way:

"I wasn't much good. I might have been. I did pretty good on the line in high school, but when I got to the Academy, I made a mistake. I decided to go into the backfield. I could run fast, and I was big and strong, so I wanted to carry the ball and be the fellow who touched it down back of the goal line. That was the mistake. I could run fast all right, but only in a straight line and all they had to do was spot the rut I was running down and put somebody into it. and I ran right into him. It always ended up with the both of us knocking each other down and the coach hollering, Swede, Swede, where's the brains for your feet?"

Just about the time the fellows down in Norfolk were thinking up adjectives to go before the name Swede, he'd knock off work and take the squadron over to Granby Street and tear all the kinks out of their minds with a party. They were not quiet parties. Everybody got up on his hind legs and roared. Swede was all for it. He wanted the fellows to feel, at the end of a night's uproar, all ragged and conscience-stricken and as if they had been away from work for a month and as if they had a lot to make up for. The fellows took their girls along.

The girls had come from all over the country to be in Norfolk with their men during these last days before the war got them. There were Missouri girls, Florida girls, Oregon and Washington and California girls, girls from Massachusetts and Louisiana and Ohio. The war wasn't biting very deeply into the country last May, but it was biting deep into the hearts of those girls as they used up the last days and minutes. What they thought of Swede at that time is one of those unmilitary secrets.

The days were going very fast. The minutes were hustling by, and, if they didn't, if some of them chanced to have any languor in them or ease or sense of timelessness, there was Swede's soft, demanding voice always at hand to crack the whip. What they think of him now need not be a secret. They must know that it was the habit of working like hell for what was wanted that saved the lives of those who were saved. Swede's wife, the former Sadie Roberts, came up from Birmingham, Alabama, with their two daughters, Melissa, then a babe in arms, and Swiss, a little less than three, and she went along on the parties. Swede calls her Missy, and everybody sang "Missyssippi" to her and "Missy you."

They were that kind of parties, very good, because everybody liked everything that anybody said or did. And John Taurman, a lad from Cincinnati, was the ladies' man of the gang, taking a kidding on that account, and Cookie (Ensign James Hill Cook of Louisiana) had just been married. He was twenty-one, the baby of the outfit, and he went around feeling sorry for every man he knew because they couldn't marry his Margie; they had to put up with second-best in the way of a wife. So, he took a kidding on that account, a "Lookie, lookie, here comes Cookie" kidding.

No, the parties were very good and very concentrated. Frenchy (Ensign E. L. Fayle) had been born in France of American parents and had lived most of his life in France. He supplied French songs on the slightest provocation and, if there happened to be a moment in which he could be heard, Lieutenant Jack Barnum of Poland, Ohio, was sure to holler, "What's all the silence about? Party, party, let's have a party."

Barnum was, in a little while, to sound his special clarion call at sea when Taurman and Frenchy, who had been thought dead, were found alive. Only ice cream was available for that party, but everybody tore into it as if it were something good. And, a little while after that, Barnum found a bottle of whiskey on Guadalcanal and came bearing it and crying, "Party, party, let's have a party."

Swede presided at that one, too. There was enough whiskey to give an ounce and a quarter to each man and the fellows made enough racket over it to make the moon go behind a cloud. When the day to leave Norfolk came, Swede sprang it on the fellows suddenly. He didn't want the goodbyes to be too long.

He knew what he was about. When a man takes off from those who love him to meet the enemy and kill those trying to kill him, he has to pull a shade down in his mind blacking out all thoughts of home. The fog of war will do it for him all right, but it's better if he helps things along himself. Then he has nothing to jitter his mind if he should be thrown up against it before the fog of war has got thick enough.

Swede didn't give his own wife any more warning than he let the other fellows give their girls. He just said, "We've done enough for here. We'll finish up with the rest out there." And said "Goodbye," and put his wife and two babies on the train to Birmingham and said, "See you soon, I hope."

The leave-taking was abrupt and there was nothing in it of pleasure. "Swede take care of yourself," said Missy, standing on the platform there alongside the train. He said he'd do the best he could in that direction.

"You're mine, too," she blurted out unexpectedly, "not only the Navy's."

He stood smiling at her awkwardly, and she thought, it was true, the Navy had had him first, he had asked for the Navy and had wanted the Navy before he had asked for her. "I hope so," he said.

The train for Birmingham started south and a little west. The planes went that way, too. They passed the train a little way out. It was going right along through that flat tidewater country there, looking small and busy as a busy boy's toy. The smoke was puffing out of it with gusty little huffs like happy breathing, and Swede thought of going low and letting the smoke from his family's train touch along the underside of his plane. But he didn't do it.

It would have been almost like a touch from Missy. But Swede was trying to pull the shade down in his mind and black out the thoughts of home, and the formation kept high, Swede leading, and the fellows never knew what train they were passing over. Then one thing went wrong with one plane and another thing with another, and the whole formation had to look around in the mountains and past the mountains for a place to sit down. They found it in Birmingham. Swede started to stomp around, hollering for quick action, keeping the men working at night, never quitting the airport once, and all the fellows looked at him as if to say, what the hell kind of a mahout is this? Here the guy is home, they told themselves, and he never even goes downtown to say hello.

Then Taurman said it to Swede. "What are you so sore for if we have to wait over a day?" he said. "Missy's train will be in by then and you'll get to see her again."

Swede looked at him startled. He had been pressing so hard to get into the war and get it over with that he had never connected up the fact that when he was in Birmingham, Alabama, he was home.

It worked out that Swede had a few minutes more with his family.

"Did you finish bombing the Japs already?" asked Swiss, and clapped her hands, glad to have her daddy back.

And that's all Swede remembers or Missy either of that second leave-taking.

In San Diego on the ship, a moment before sailing to the war, Swede got free and put in a telephone call for Missy.

"Hurry, please," he said. He heard his home phone ring once. Then the operator broke in. "I'm sorry, sir," she said, "but we have to disconnect." It was sailing time. "Can't you give me one minute?" "I'm sorr ..." Then the operator's voice ended.

Missy heard the telephone ring. She was two rooms away from it. She heard three rings altogether. On the second ring, she knew suddenly that this was Swede and began to run. She picked up the telephone just as the third ring ended. There was only silence coming from it. Swede was sitting more than two thousand miles away with a disconnected telephone in his hand, struggling to pull down the shade in his mind and wondering uneasily if this piece of bad luck meant his luck would be bad all the way now.

"Bullet" Bob Feller, 1936 Goudey Gum Baseball Card

2 - SWEDE LARSEN PICKS SIX

THE FIRST THING THE FELLOWS learned when they got into Honolulu from Norfolk with their new planes was that they were too late. They had not been expected that soon, but even so they were too late.

The rest of Torpedo 8 had not been able to wait, but had gone out to sea on the Hornet with their old planes in an awful hurry, hotfooting it for some kind of a brew-up out there. Then Swede and the fellows he had in charge had to sit around until they got orders. This waiting around for flag plot [tactical decisions, from the Admiral's flagship] and air plot [Air Operations Control on board the carrier] to make up their minds what to do with them is one of the nervous things that is always happening to airmen in the Pacific.

It's natural to the war out there, for the airplanes in the Pacific, warring from island bases or ships with limited facilities against island bases or ships with equally limited facilities, have to make a one-punch fight out of it. They are the Pacific's Sunday punch, and if they don't get home everybody starts looking around for flowers to remember himself by.

To understand that, look at it in this way. Here's a guy coming in to knock your block off with a Sunday punch. If you ward off that punch, he'll throw another and another. So, the thing you have to do is knock his block off. Then, he will stop throwing punches at you.

Now, you can't throw your punch to knock his block off until he gets in range. But if he is in range of you, you are in range of him, so while you are throwing, he is throwing, too. This would not be too bad in the kind of defensive-offensive war we were fighting in the Pacific in 1942, except that the fighting in the latter portion of the year took place along the fringes of the power of both sides.

We had to go farther to get at the Japs than they had to get at us, true, but not so much that it changed the general picture of both sides having to fight off ships or bases that could handle only so many planes and no more.

That is, neither side could ever put into action at one place at one time enough planes to take care of all the work that had to be done; warding off the enemy's Sunday punch while, at the same time, knocking his block off with your own.

So, when either side threw his Sunday punch, he had at the same, time to lower his guard, as they say; at least lower it to a certain extent.

That is, the planes thrown at the enemy's ship or base are planes stripped from the job of warding off the enemy's assault. In circumstances like these, timing can be the determining factor, and if your clock is off, you're like a fellow running to catch a train that left an hour before. You'd be better off not doing it. In the battles in the Pacific, when we threw a punch, we never could count on getting a chance to throw another. We might get our blocks knocked off before there was time.

When the planes took off from home, they never could be sure that home would be left alive by the time they got back. If you are the underdog, and we were invariably the underdog in the Pacific in 1942 all the way from [the loss of] Bataan [in Luzon, Philippines] to Guadalcanal, even when making the show, even when forcing the attack, you may choose to hold your punch until the enemy has landed his. You hope to use your planes for the dual job of first protecting your chin and then later knocking the other guy's block off.

But you are likely to find your planes overwhelmed by the enemy's punch and you buried under them and they without enough strength left to knock even their own knees. If you throw your punch too soon, you are likely to find the enemy so far off that the planes don't have enough gas to make a round trip out of the job or to maneuver successfully for the assault.

Or they are likely to find the enemy with his guard up so that the attack never can be rammed in with full force. Then it's you who take the ramming, particularly if the enemy has been ungracious enough to refrain from making your mistake.

Too soon is as bad as too late. But the timetable of an attack on the Pacific is not only a question of when. It is also a question of where, for, in dealing with moving objects such as ships and planes, timing becomes a blend of place and time. Not only must you decide when the enemy is going to throw his punch, but from where, so that you can move your ships around to have your maximum strength in range to hit back. Where and when he is going to hit you and where and when is best for hitting him, how much is he going to hit you with and how much is he going to protect himself with, how much should you leave at home and how much throw into the punch; these are the factors that come into the timetable of battle on the Pacific and they weigh quite a good deal on the staff mind.

They would on yours, too, if you knew your first punch had to be the right one or there would be no more, and if you knew that once you made the decision that was the end of it, you had to ride it, to victory or to the grave.

So, the fellows from Torpedo 8 sat around waiting, wondering what was cooking and who it would be that got cooked. Then Admiral Noyes* called in Swede Larsen and said he wanted six bombers to go to Midway and work out of there.

*Vice Admiral Leigh Noyes (1885-1961), Commander, Carrier Division 3.

"Me and five others," Swede said.

"I think not," the Admiral replied. "I think you'd better stay here with the rest of the men and the planes."

Swede thought many things, but he said "Yes, sir," and walked away with that old-line sergeant-major walk of his and summoned the fellows to give them the news. June was almost there then.

June in Honolulu is very special, like April in Paris or November in New York. But that June did not get past the gates at Pearl Harbor and did not get into the bare, bleak, raw-smelling room at CASU (Carrier Aircraft Service Unit), where the fellows sat to hear what Swede had to say.

Swede stood in front of the room with a blackboard behind him and some maps. He didn't use the maps. He just said there was a lot of weather out on the Pacific, a whole front, and the Japs were hiding under it, coming along east with it, and Admiral Noyes had asked him to send six of the new planes to Midway to wait there for whatever happened.

"The Japs are not coming for the ride," Swede said, "so something is pretty sure to happen. Maybe they're going for Midway or maybe they'll bypass Midway and swing around and come right for Hawaii. Either way, it looks as if land-based planes on Midway will have a chance to get in there and make some noise for themselves. And there will be noise made at them. That's a point to consider, the noise made at them.

"Well, so six planes are going, and the rest are staying here. That's the orders. I don't like to have to pick out the six myself. I'd rather have volunteers because I think there's going to be a battle and there's liable to be a lot to it. I don't think the Japs would come all this way just to throw confetti."

The fellows didn't think so either. The talk was made in the afternoon.

"You can hand in your names in the morning," said Swede. "That'll give you a chance to sleep on it and I'll pick the six who go, out of the names."

He stopped talking suddenly and decided to have another go at Admiral Noyes. This was his show. They were his planes. He had put them into commission for combat and he ought to be with them when they got in their first shots.

The older people of the United States can be content with the young men they have produced and who are now away from home fighting the war. Most of these young men are astonishingly practical, very realistic and hard-headed in the midst of all the desperations of battle, and when they are asked to volunteer for a particularly dangerous mission, they do not become romantics, rushing heedlessly to die for a dear old Rutgers.

They examine the situation coldly, all the ins and outs of it, and independently of their officers arrive at decisions which are eminently satisfactory to those concerned both with the prosecution of the war and the nature of the Americans fighting it. They do not let the war down, nor do they let themselves down. They are altogether very sound, good people.

On several occasions, I was present on battlegrounds in the Pacific when requests for volunteers were made. There was always a pause for thought and nobody looked at anybody else, but looked only inside himself.

You could see on the faces of the men concerned the reflections of a debate going on inside their minds; a grisly kind of arithmetic that has never been taught in our schools, but which our men seem to know well, a private adding up of profit and loss, of profit to the war and their particular role in it and of loss to the war and their particular role in it.

I saw a young combat reporter volunteer to climb a naked ridge in the face of Jap sniper fire and kill the snipers there when it was not his job to do so. The rest of the men were slow in making up their minds that the job was worth doing. They had been in action, as the combat reporter had been, a long, long time and were tired and were safe where they were. They did not see any reason for rush about the snipers.

They were willing to wait to let the snipers expose themselves and pot them from a safe distance. The combat reporter did not see any military reason for rush either, but he volunteered for the job, anyway, on the very practical grounds that I was there as a correspondent for many newspapers, that I might not understand why volunteers were slow in forthcoming, that his job as historian of the unit to which he was attached involved putting the reputation of the unit in its proper light.

So off he went, ducking away very swiftly, hoping I would believe that a regular had volunteered and not he, and, when I caught up with him, he was diffident about what he had done and said, "I broke the rule of good conduct, which is never volunteer for nothing."

Our men volunteer for suicidal missions, not to seem brave, not to win medals or promotion, but only when, independently of their officers, they decide the possible gain is worth the probable loss. They decline to volunteer only when their private arithmetic does not add up that way. They do things that seem crazy, but never for crazy reasons, almost always for reasons that stand the test of sound adult thought.

I once saw several hundred young men, none of them yet thirty years old, all veterans of hard fighting, nearly fifty of them winners of medals for heroism beyond the call of duty, fail to produce a single volunteer because each, consulting himself privately, had arrived at the same result: the possible gain was not worth the probable loss.

A desk officer, unfamiliar with a certain type of weapon under actual combat conditions, had thought up a use for that weapon which looked good on paper and wanted it tested in a real fight. The commanding officer of the unit fighting with that particular weapon believed the new idea impractical, but the desk officer outranked him substantially and he said he'd put it up to his men.

Volunteers were to write their names on a bulletin board at the command post. No names were written the first day and none the second. The commanding officer took down the bulletin and said to his men: "I guess you guys are all going chicken on me. That's what I'll have to report back."

I remember more than a dozen young fellows spoke up with, "No, sir," and "That's not so, sir," at the same time. Then they explained: "It's just that we're doing all right doing the way we're doing, and this new stunt is just crazy. Not only would nobody live through it, but you just wouldn't do any good with it. You'd be killed before you could do any good."

"That's what I think, too," replied the commanding officer, "but I just wanted to know what you guys were thinking and make sure you weren't chickening out on me."

Then he went back to report to his superior and there something equally heartening happened. The desk officer accepted the verdict of the men without question. He was democratic in his mind, too, American in his mind, and had found out that American soldiers in battle arrive honestly at honest conclusions.

"I guess we'd better forget it," he said. "I guess it was just one of those pipe-dreams that come to a man who has to stick at a desk and can't go out and see for himself."

Contrast this with the fascist notions of the Japanese soldiers, who cannot conceive of their superiors ever being wrong and continually try to carry out pipe-dreams and invariably go up in smoke. Their desk-dreamers do not trust the judgment of their subordinates and many, many thousands of Japanese lives have been wasted on that account alone. The democratic process with the habits of thought it has produced in our fighting men, has given us a valuable, life-saving system of checks and balances in this war.

When Swede Larsen asked Torpedo 8 for six crews to volunteer to standby at Midway for the Japs coming in and gave the fellows a night to sleep on it, the process of deadly arithmetic began in the minds of everybody concerned. In this case, they all knew it was their duty to go, but they all knew, too, that the men in the squadron had been in training for varying lengths of time, anywhere from two months to eight months. Some felt they were more ready for a fight than others in the squadron and some felt they were less ready, and that, as it turned out the next morning, proved the determining factor. Not desire to show up as a hero, not a restless, reckless itch for action, not curiosity to see what a battle really looked like, not heedless lust to kill Japs.

All the men wanted to show up as heroes, were restless for action, were curious to see a battle, wanted to kill Japs, but the volunteers the next morning were in almost every case the men who had had the longest training, and in that fact could be seen that the arithmetic of young Americans was good. It was practical and sound and evaluated a situation in an unemotional, matter-of-fact way.

Then it was Swede's turn to be hard-headed about the thing. He had asked Admiral Noyes a second time to let him go to Midway and had been turned down a second time. But he had had longer training than anybody in the squadron and he felt further that it was his place to be there. The third time he went to Admiral Noyes, the Admiral was very curt, and Swede had to say, "Yes, sir, very good, sir," in a stiff way and keep his thoughts to himself in a very stiff way.

Then he went back to his bunk in BOQ (Bachelor Officer's Quarters) and picked the six crews to go from the long list of volunteers he had. He picked them unemotionally. Two were picked because, while they were good fliers, they were not as expert as others in landing and taking off from carriers, and there was a chance if the battle went badly the part of the squadron left in Pearl Harbor would join it from a carrier. Two others were picked because they were very close friends and worked well together. The remaining two got to go because they had had the longest training.

The six planes took off on the Tuesday morning of June 2. Everybody was down to see them off. It was like a college football team going off for the big game. That is one thing that strikes everybody who has the privilege of watching American young men carry on the adult business of war; they are adult when that is what they have to be, and all the rest of the time they act their age.

There was a football type of kidding going on—"Don't forget, when you drop, to open the bomb-bay doors," that kind of thing—and there was cheering and envy and one of the fellows was crying. That was little Billy Bragg, who came out of one of the Carolinas looking about fourteen years old and wanted in the worst way to see what a battle looked like. He had been after Swede ten times a day to let him go, but he hadn't got to go because there were other men better trained. The flight to Midway was long and difficult and very hazardous. It was like hitting a dime from a mile away with a good rifle, but with a bad crosswind blowing.

All the planes made it in fine shape. They hit the dime on the bullseye, and when Swede got the news, he felt happy, really happy inside.

That was Tuesday night. The fellows of Torpedo 8 on Midway spent Wednesday, the last full day of their lives, waiting for air plot to make up its mind, filling in the time by checking their planes. And Thursday morning all but two were dead, or, as the Navy says when the bodies cannot be found, "missing."

The 'Lady Lex' USS Lexington (CV-2) underway at San Diego, CA, October 14, 1941

Thick smoke rises from an accidental explosion on board the Lexington which led to the decision to scuttle her on 8 May 1942

3 - THE BATTLE OF MIDWAY

THE MEN OF TORPEDO 8 who were flying the old planes took off from the Hornet at sea. The crews in the six new planes—the Grumman Avenger—took off from Midway. They were all in the air heading for the Japs before the taste of morning coffee was well out of their mouths. That was Thursday, June 4. The fellows at Midway were dug out of their sacks at five o'clock in the morning.

"Holy crock, Doc," and so forth, "Grab your sock. It's five o'clock, Doc," and so forth, "Time to get up."

Third-class ordnance man Lyonal J. Orgeron stirred no more than a log might through the bellowing. He was replacing William Lawson Coffey as tunnel gunner in one of the planes. Coffey was a top-grade mechanic and it had been thought more sensible to let him work on the planes instead of with them.

"Let him sleep," whispered Coffey, "I'll go."

Where the bellowing had failed to disturb Orgeron, the whisper didn't. He snapped awake instantly.

"Hell, no!" he cried, slapping his feet on the deck, "this is my pigeon."

He looked around the tent and saw most of the fellows had gone and grabbed up his shoes and socks and shirt and ran barefoot to the airfield, carrying his shoes in his hand and crying, "Wait for baby."

Back in Pearl Harbor, the men who had been left in reserve knew there was a battle going on and that Torpedo 8 was in it and that there was a lot to it. That's all they knew Wednesday, Thursday, Friday, and most of Saturday.

The communications channels were jammed up with priority messages. In a situation like that, what has happened doesn't rate at all. It's the immediate future that counts and the past has to take its place at the end of the line.

Like most bomber pilots, torpedo bombers are usually big men who can't fit into the narrow cockpits of fighter planes and they are calm, and they don't worry until they have to.

That's the way Swede Larsen looked those days and all the others, too, big, for the most part, slow-moving and calm-seeming. But the calmness was just something they were hiding behind.

They knew their business; going in fast and low and drawing their planes across the mouths of the enemy guns, as if their planes were handkerchiefs wiping off those foaming, frantically chattering mouths, and knew that when their number was called there were not often any two ways about it. It was not often that a torpedo plane had a chance to be shot up or shot down with the crew getting out all right.

In those days, when a torpedo plane got it, it usually took it head-on, the torpedo plane and the shell with the plane's number on it diving head-on into each other, and all you saw out of that was flames with maybe little black streaks of metal flying up and out.

The scuttlebutt, as the Navy calls rumors, was flying around pretty thick in those days. Some of the rumors were good and some bad. Swede Larsen stuck close to the command center, and on Saturday afternoon a commander in Naval Air Base Defense gave him the score on Torpedo 8.

One plane had come back, he said. Swede spent a moment taking that in.

"Whose was it?" he asked at last.

Earnest's, the Commander told him. This was Albert K. Earnest, a tall, dark-haired boy from Richmond, Virginia. The Commander was very busy. The Battle of Midway had cooked up to a boil.

"I appreciate your taking time out to tell me this," said Swede.

The Commander looked at him strangely. "Well," he replied, "the others are all down somewhere. We don't know where they are, but we're looking and hoping."

"Yes, sir," said Swede. "Thank you, sir."

Swede doesn't remember much of what happened the rest of that day. He remembers he felt mad at the Japs. That was his first emotion, to get out and get after them. All the other emotions he had that day do not seem to have counted enough with him to remember them. The story of the attack came to the fellows gradually. It had been a hands-down situation.

The Japs had been a little over a hundred and sixty miles away. Everything in the situation was made to order for them, but it was a question of hitting them then, no matter what the conditions, or of not hitting them at all, so Torpedo 8 and some Army B-25's following them, hit them.

The Japs had anti-aircraft and something between twenty and thirty Zeros waiting for the attack. The weather was as if the Japs had thought it up.

There was a low overcast and our fellows had to come in under the overcast and hustle along a rut less than a mile wide. That was all they had to maneuver in; less than a mile. It may sound like a lot to the infantry or artillery, but a mile is about twelve seconds to an airplane.

All the Japs had to do was park themselves along that rut and wait and the farthest away they'd be from any of our planes was twelve seconds. That's what they did, parked their Zeros along that rut and poured their anti-aircraft fire down that rut, and our fellows, going in, got it from all sides, from top and bottom and head-on.

Our bombers had no help at all; no fighters to break up the Zeros, no dive-bombers to divide and thin out the anti-aircraft fire. That was the way the thing shaped up. But it was hands down there.

It was something that had to be done, and Torpedo 8 and the Army B-25s went ahead and did it, stuck their chins out and followed their chins on down along that terrible rut, and only Earnest's plane came out the other end.

Guns were working and the big, vicious, bull-whip cracks of anti-aircraft fire were smacking all over the place. Earnest was trying to line up his target.

The turret gunner cried, "One down, Zero down," into the interphone, and cried some other things, but the chugging of the machine guns made it impossible to hear him. Then, suddenly, his voice was very clear: "One of ours down, no theirs, theirs, the fourth of theirs down, hitting now. Goodbye him. He's hitting like a ton of bricks."

Those were the last words he was heard to say, for right after that Earnest's plane got belted and the anti-aircraft fire seemed to smack down along the length of the plane inside, and the plane lifted into the air crazily as if it had been kicked, and shied violently like a frightened colt and began to fall.

It couldn't jink anymore and it couldn't corkscrew. That meant, for as long as it sat up there in the air, it would sit like a tin duck in a gallery.

The air in the plane felt all smashed and smelled metallic. Manning, the turret gunner, was dead. The tunnel gunner, Harry Ferrier, had got hit in the head and was fighting dazedly to remain conscious. The whole instrument panel had been shot away from right in front of Earnest. His face felt torn open, but his mind kept ticking on steadily. He saw he was heading toward a cruiser and he just kept going that way, falling lower and lower, struggling to keep the plane level so that the torpedo, if he lived long enough to drop it from where it would hit, would launch properly.

Earnest was twenty-five years old, but that's all he kept thinking of at the time; keeping the plane level so that the torpedo would launch properly and hoping to get close enough for the torpedo to have a chance to hit.

The plane kept taking bullets all the time. There was no way to dodge them.

But, by some miracle, it was all small stuff, and the plane kept right on going and Earnest's mind kept right on ticking and Ferrier kept fighting sluggishly, in a dazed way, to keep conscious and work the tunnel gun.

Then Earnest dropped the torpedo and found himself past the Jap ships. He knew he was past them because the anti-aircraft fire wasn't coming in on his face anymore. It was on his tail, and he looked around and saw that two Zeros were there, too, out to finish him off. He was very tired, and he didn't care much. There had been a big letdown in him after the torpedo had got off. Besides, there wasn't much he could do about the Zeros. All his controls had been shot away.

He was flying on his tabs. The tabs are the things that control the settings of the elevators and rudders and all the flight controls. It was as if the steering wheel of his car had been shot off and he was trying to steer with brakes. The Zeros didn't get Earnest. They tried for about twenty miles.

Then they ran out of something, neither Earnest nor Ferrier know what, gas, or bullets, or maybe they just ran out of guts. Anyway, they went back to their ships and left the crippled plane and the wounded men alone to the sea and the sky. The sea and sky were empty. All the fire was out of them now and they were very quiet and peaceful. The plane was still flying. Earnest's mind was still working. With the pressure off him to keep conscious, Ferrier gave up and became unconscious for long periods at a time.

But Earnest couldn't rest. He didn't allow himself to hope he'd get back all right, but he thought he owed it to the plane to keep it going as long as it showed any inclination to.

Flying on the tabs was tedious and laborious work. When a man gets through a battle, he wants to fall asleep or let out a roar or do something to give off steam. To have to throw himself immediately into tedious drudgery of the sweatingest kind is really galling. It takes all the nerves in a man and really galls them in a low-down, scraping way. It is much more strain than fighting.

But Earnest stuck to it patiently, drove himself to it, and got back with the dead man and with Ferrier. A funny thing was that, while his mind worked well on the problems of the job he had to do, it worked very sluggishly on everything else.

He had no instruments left with which to find his way home. He just had to head down a bearing of the sun. But he hit Midway on the nose all the same with a job of dead-reckoning navigation that was first class. While he was doing that, he was trying to figure out about himself.

He was covered with blood from the wound in his cheek and had been knocked around and bruised so much by the up-and-down whippings and shyings of the plane that he felt as if he had been hit in more places than the one.

He felt riddled and, as a matter of fact, he thought for quite a while that he was dead. Then he figured out sluggishly, while the plane droned ahead and the laborious drudge work piled on, that he couldn't be dead, he must only be dying.

He kept listening to himself and trying to feel himself with his mind, wanting to know if he was getting weaker, but he couldn't figure out anything about himself. His mind worked too slowly in that direction and after a while he decided, slowly, "the hell with it," and stopped thinking about anything at all except the work he was doing.

When the plane was twenty-five miles away from Midway, Earnest saw big columns of smoke standing up to the sky and knew he was home. It looked to him as if die Japs had sunk the island. They hadn't, but that's the way it looked.

When he let down his wheels for landing, he found that only one would go down. The other was stuck. He decided to slide in on the belly of the plane. Then he discovered that the wheel he had let down wouldn't come up. The hydraulics had been shot up and the hand apparatus, too.

So, there he was, after as desperately fought an engagement as any in the history of United States fighting men, after as laborious a flight home as can be made—there he was, faced with the problem of a landing on one wheel, the stickiest kind of landing known.

But he did that, too, all right, landed nice and easy on the one wheel and let down the other side nice and easy. When Coffey and the other mechanics and the crash wagon and the meat wagon, as the fellows call the ambulance, gathered around, nobody could figure out how the plane had stayed up, except that there were so many holes in it that maybe it had become lighter than air. But two weeks later, Earnest and Ferrier were back flying with Torpedo 8, and when they saw Swede Larsen's memorandum changing the squadron's slogan from 'Attack' to 'Attack and Vengeance,' they each said, "That's where I come in."

Then Missy, sitting home in Birmingham with her baby daughters, started a letter to her husband. The newspapers she had been reading were describing torpedo bombers as the American answer to Jap suicide squads and to Nazi elite troops who ran singing into machine-gun fire from which they had no protection.

"Can you imagine poor dumb me," Missy wrote Swede. "Here I was thinking you were as safe as in a church and then I find out you're in the middle of a mad house."

American women ordinarily get from their men at war a feeling of confidence in the machines with which their men are working. The women of fighter pilots, for instance, and fighter pilots themselves, think that piloting a fighter plane—those whippy, snorting little platforms that carry guns up to the mouths of other guns at four hundred miles an hour—is the best and safest job in the war.

Soundly trained tank-destroyer men are positive no tank can stand up against them. And soundly trained tank-corps men shiver at the idea of having to fight tanks and feel very sorry for their friends in the tank-destroyer battalions. And the women of tank-corps men feel glad their men are not in the tank destroyers. These are natural feelings that arise from something very deep in Americans.

Good morale is only a small part of it. The major part of it is an understanding of their machines, an understanding so complete that the men are able to communicate it to their women.

For American war machines are all able to do not only the jobs of destruction for which they are designed, but the jobs of conservation, too. American inventors, designers, manufacturers, military men, in fact, every last man concerned with the making and using of our machines, has insisted on the same thing—that an American war machine must not merely be able to destroy the enemy, but must also be able to conserve the life of the Americans using it.

Their insistence is not because of some legislation somewhere that can be put aside in moments of stress or evaded when nobody is looking. It is, instead, fundamental in us, a part of our temperaments, a natural habit of thought growing out of the democratic process which is the American way of life.

So, when the news came through of the destruction by the Japs of all but one of Torpedo 8's striking force of machines, there was stagger and fear in the mind of Missy, and in the minds, too, of many of the surviving men of the squadron.

At the beginning, there was no time for anybody to do anything about it.

The Jap was running away and had to be pursued, and the remnants of Torpedo 8 were gathered up; all of them, the scared ones and the desperate ones and the fellows who just shrugged it off and thought, "Hell, dying is something that happens to everybody once."

They were all thrown into the pursuit. Torpedo 8 had no luck with their end of the chase. They were where the Jap refrained from going. And in this unhappy time of straining for action and of getting no action, the thoughts kept cooking in them, and they had, to keep them going on through their thoughts, only a sense of obligation toward the war; a feeling for their "duty," as it is called, to their country.

That, it turned out, was enough for the time being. It was enough at home and enough on the battle line. Missy, as a woman, wanted her husband. That was in the letter she had written. As a mother she wanted her children to have their father. That was in the letter, too. But she understood that torpedo planes and Swede had been wedded to each other by war until death or the Navy did them part. Taking that into account and taking into account her need to do her duty to her country, she decided against doing anything that would harass her husband and impede victory in the war.

She had written the letter out of the terror in her and, out of a sense of obligation to the war, she controlled her terror and tore up the letter before mailing it. There were other letters of a similar nature written or planned by Torpedo 8's women. None that I can find out about were ever mailed.

The men on the battle line came to the same decision in their own way. There was nobody to coax them into it. In fact, it developed, with the best intentions in the world, that the pull was all the other way. Their friends in the fleet kept coming up to them and saying, "Gosh, I'm glad to see you, I thought you were dead." The temptation was unavoidable to reply, "No, not yet."

The replacements weren't very happy either about being picked to fill the holes made by the Japs in Torpedo 8, and there was talk all around, in the fleet, in the newspapers, among the civilian experts, among even the men themselves, about the plane not being good enough, and about this being needed and that being needed and so forth and so on, a real groundswell of discontent.

The men of Torpedo 8 rode the groundswell and kept on going through it. They were not unreal people facing an unreal situation. The situation was very real, and they were human. That is one of the magnificent things about what happened later, that they were all quite human people, exactly the kind the rest of us are, and did what they did despite that. So, in the meantime, they bucked a little in the groundswell, but kept on going through it.

However, Swede was dissatisfied with the situation. He had been moved up to the late Lieutenant Commander Waldron's place as commanding officer of the squadron. He did not want his men to have to whip themselves to their work by their sense of duty. He knew duty would make the men fight, but he did not believe it could make them fight efficiently. For a man to best an enemy as resourceful, heedlessly tenacious, and implacable as the Jap had proved himself to be, he must have his wits about him. Desperation was not enough. Not only must his heart be on fire, but his nerves must be firm and in order.

Swede was not an orator, or a psychologist, or a skillful, tricky manipulator of men, able to prove brilliantly to the less clever that black was white.

He was just a fellow who felt one way about something while the fellows around him seemed to feel another way, and he thought if he explained how he felt, then the other fellows would be sure to see that he was right and would feel the same way he did. It was the old-time cracker-barrel way of doing things. It implied a faith in your fellow man which is part of the democratic process, and it turned out that that faith was the best; better than any cleverness, or skill, or fund of learning.

Lieutenant George Flinn lined the men up on the deck of the Hornet at sea.

Flinn is a boy from a wealthy Pittsburgh family. Dave Hammond stood near him. He's a lad who has spent all his life working with his hands. Red Doggett was there, too, a cheerful, talkative enlisted man with enough drive in him to get him off the deck and into the air as a pilot.

The South was there, the North, the East, the West; the rich, poor, and in-between; college men, CIO men, AFL men, farm boys who had never got past grammar school, salesmen, lawyers, clerks, truck drivers; Bruce Harwood, a fighter pilot who had been asked by Swede to come into Torpedo 8 and had said, "Well, I don't like the idea, but if I have to I will," and had been told by Swede, "Yes, you have to," because Swede was sure he'd be good at it.

There were about a hundred and fifty fellows there, some eighteen years old and some fifty years old, standing in rows, shoulder to shoulder, to hear what Swede had on his mind.

"Torpedo 8 still exists," said Swede, "and the way I feel, it exists to get vengeance. That's what I see is our job now, revenge."

Then he pointed out, what everybody there knew, that it was a "rugged" war.

"A man has to be able in it and lucky," he said. "Nobody can do anything about his luck except ride along with it and hope for the best, but everybody can do something about being able. And my experience is, when a man is able, he has a better than even chance to make his own luck. Getting able is just a question of working at it, and that's what we're going to do, work, work, and work, and then work some more. I'm not going to make myself popular with you. I'm just going to make you work.

"And then we're going to hit the Japs wherever we can. We're going to do what the situation requires us to do. If it's a hands-down situation, we're going to put our hands down. But the situations are all going to be good for us because they're made to order for our job, which is vengeance; which is to blow into the heads of those Japs who pulled the trigger on us at Midway the realization that in the long run they did the sorriest, stupidest, worst job for their country that they could have done. Their country would have been better off if they hadn't been born. That's our job, to teach them that and teach their country that.

"We've got a good plane for the job. I know there has been some excitement around here about the plane. You can see for yourself, just by looking around and listening, that the less experienced a man is with the plane, the more prone he is to get excited about the way it works, and the more experienced a man is, the better he thinks the plane is. That's what I see, and you can find out very easily whether I'm wrong. When the men on the ground working on it are good and the men who take it up into the air are good, then it's a very good airplane. It has a good name: The Avenger. We've got a good job: Revenge.

"That's what I want, vengeance, and I think you want it, too, and think that if you stop to figure it out you'll be glad, too, as I am, that the Navy is putting us where we can go to work and get it. We've got a score to settle and a chance to settle it. Who can ask for more?"

Those are not the exact words Swede used. Swede doesn't remember his exact words, and nobody thought to mark them down. All Swede remembers is that he told the fellows what he felt, and this was the gist of it. Whatever the words were, they seemed to have been the right ones. If Swede had told the fellows what they didn't believe or couldn't believe or didn't want to believe, then nothing would have happened.

But something did happen. They all, from that day on, committed themselves to vengeance. They stepped with their minds out of their old familiar world and into a world in which their dead could live with them fiercely. Revenge was their job. When they got to Guadalcanal, they worked at it relentlessly.

VT-8's Navy Grumman TBF-1 "Avenger"

4 - A Day at Guadalcanal

THE FIRST FLIGHT OF AVENGERS of Torpedo 8 took to the air over Guadalcanal in the black of the historic morning of August 7. There were no Jap ships around for them to torpedo, so they were carrying bombs. Everybody on board the carrier knew how anxious Torpedo 8 was to get at the Japs, so there had been kidding to the effect that they wouldn't get any.

The kidding took the line that when zero hour came or, as it was called, D-hour, H-day, and Torpedo 8 would find no gas in its planes, or no engines, or no propellers, or no something or other. Anyway, it wouldn't even be able to take off.

This had made Lieutenant Dewitt Peterkin* and Clyde Hammond, whose responsibility it was to have the planes ready, climb all the way out on a limb and say that not only would every single plane in the squadron be fit to take the air, but they'd bet on it. A number of crafty businessmen on the carrier had heard of this brash and heated statement and had roped off Peterkin and Hammond and had made them put up the money.

*Engineer Officer Dewitt "Pete" Peterkin.

So, there was considerable interest in the takeoff, and, as the planes hurtled off one by one, the businessmen were advising Peterkin and Hammond not to count their chickens before they were hatched, and Peterkin and Hammond were counting them. Swede, who was leading the first flight, took off last.

As he gunned his motor and let her go down the deck, a great gout of oil shot out of it and caught the businessmen, the engineering crew, and the plane handlers in their respective kissers. But Swede was hot to go, so he just kept on going. The oil was all over his windshield and he couldn't see anything except oil, but there was no ignition switch on his mind that he could turn and make himself stop going.

Torpedo 8 had arranged to rendezvous in the sky, circling five minutes and then going on, leaving behind anybody who had not showed up by that time. Swede had not showed up, so the rest went on, Barnum acting as flight leader.

It was a soot-black, rainy, blowy morning, and Swede was in the middle of it, listening intently to his motor, hoping that whatever it was that was shooting oil would stop, and, when it stopped, wondering if he had enough oil left to keep his motor from grinding itself up.

The indicator showed there was enough and Swede thought, "Indicator, if you're lying, that's the last thing you'll ever do," which, as a matter of fact, was correct, since, if it were lying, it and everybody else would wind up in the water.

In the meantime, he kept slewing his plane around to look out the sides, hoping he'd catch a glimpse of the others. Once he did and went chasing after them, but they had gone somewhere else and he couldn't see where that was because of the oil. Nobody was using his radio then. That had been the orders; radio silence until the Japs were hit and made a noise. Swede knew where the target was, and figured to go on to it alone, and maybe get there before the others had finished it off.

A torpedo bomber flying alone is a fox running in second place when fighter planes jump him, but that's a chance a man takes when he is hot to get somewhere and doesn't mind the manner of his going, and, anyway, there were plenty of clouds around.

Swede figured, if he saw Zeros, he could jump into a cloud and wrap himself up in it, although how he expected to see Zeros with oil all over his windshield is something he forgot to take into account. Then the oil started to clear off his windshield. The gusty wind blew fierce little squalls of rain against it. The rain cut like knives. Dawn was starting to scoop out the darkness, and radios opened up in the air all around.

"There's the objective. Hit it," Swede heard suddenly in the soft, high voice of the American South, and the loneliness went all out of the foreign air around him and he felt at home and as if this were the start of a good morning at home.

Torpedo 8's objective was at the eastern end of Florida Island across the basin of water from Guadalcanal; that same basin which, in a little while, the Navy would learn to call "Iron Bottom Bay" and the Marines "Sleepless Lagoon."

The radio crackled constantly now in Swede's ears. The Orange Commander and Red Commander were having quite a talk.

"I'll take him," said a Midwestern voice, and then there was a whole stream of blue language followed by the explanation, "I told you to watch out for a tail-end, Charlie, unless you got eyes on the back of your head," and the answer in a quiet but heavy-breathed voice: "That's where that guy almost knocked my eyes to, to the back of my head."

Warships were walking a barrage up and down the beach and somebody was telling them to take smaller steps and walk more slowly. Swede could see the muzzle flashes through the thinning darkness and could see bombs falling, looking like matches striking suddenly and flaring suddenly, and then he saw his own planes, "Hey, Barney," he called over the radio, "this is Swede. I'm on your starboard side. Move over."

The first enemy soil Torpedo 8 saw was the east coast of Florida Island, as thick with matted, bristly, woolly-looking jungle foliage as a Fiji Islander's head is with hair. But there was no enemy there, at least none they could see, only some native huts, and Swede told the fellows to save their bombs for where they could see what they were doing, and led them on to Point Purvis.

They found a Jap headquarters there and Swede waggled to indicate that here it was, the time had come to kill. The fellows didn't want to miss. They all went in very low in a type of bombing known as "flat-hatting" and dropped one bomb to a pass. Dropping on an enemy is not like dropping in practice, and they were all nervous and knew they were nervous, and wanted to give themselves a lot of shots to make sure they'd get a hit. If they dropped a stick and that stick missed, then the show was over. But dropping one to a pass gave them a chance to correct for nervousness. Of course, the more passes they made, the more chances the Japs had to get them. But the fellows weren't thinking of being got by Japs, only of getting Japs.

Swede picked out the biggest house there and went for that. He wanted the boss. As he came in, he saw little black blobs of people scurrying and saw the tracers of his machine guns taking out after them. Gunnery was Swede's first love and he had worked his men hard on that because that was their best defense. The tracers were taking out in the right direction.

The cones of bullets were whipsawing after them like swarms of eager bees. Then the boss house was set up in front of him and he dropped on it and there was a concussion that caught his plane in the tail like a vigorously placed boot. It lifted the tail up and slewed the plane around.

He fought it back to level and looked around. The boss house was still standing. So, Swede went back to it and this time dropped one right where the chicken lays the egg, and when he looked around there was no more boss house to see. Then he sashayed about a bit, looking for more work to do, found it and did it.

That's the way all the fellows spent their time during their visit to the Jap headquarters, sashaying around looking for work, and finding it and doing it. When they missed on a pass, they made another pass, and, finally, when half the bombs were expended and nobody could find any more work to do, they all set out for Marmasike Passage and Takataka Bay on Malaita Island. That was Number 2 on the list of calls to be made that day.

But big black thunderheads lay between Marmasike Passage and Point Purvis. Thunderheads are the wind's boiler factories. They are big black vastly spreading things that sit on the water and stretch up to as high as fifty thousand feet, and everybody who flies the Pacific has learned to treat them with caution. They have thermals in them that can break a plane into pieces, the way a forge hammer might, only more quickly.

So, Swede moved around to the western edge of Malaita toward Coleridge Bay, and, much to the regret of the "so solly" Japs, stumbled over Langalanga Harbor. There were little moorings there that nobody except the Japs had known anything about, preformed areas for storing boats, a protected lagoon for docking boats and plenty of floats. It looked like an excellent set-up for torpedo boats or maybe something even bigger, and Torpedo 8 looked it over carefully for a few minutes, and then went in and messed it up.

Again, Swede picked the biggest hut there. The nervousness was out of him now and he got a hit on his first pass. There was one big boom and then another even bigger one, and great gaseous-looking billows of black smoke started to pour out of the wreckage, with flames licking at the edges. That meant gasoline or maybe ammunition. So, Torpedo 8 really went to work on that place and made a mess of it.

Then, out of bombs, they headed back for the carrier. The fellows were tired by this time and flying ragged formation. Swede looked around vexed. Just then words came over the air that Zeros were humdinging along looking for a scrap, and the fellows rushed to close in a formation so tight you could walk from wing to wing. They had rushed so fast that Swede had to laugh. He looked around him and saw John Taurman grinning at him and Red Doggett waving and grinning.

They were enjoying the joke, too. But the Zeros got all the scrap they wanted before coming even within sight of the torpedo bombers, and the fellows made the carrier without having to use their machine guns.

In the meantime, Bruce Harwood had led a flight of Torpedo 8 bombers to work over Guadalcanal itself. They were given the same sort of targets; whatever they could find worth a bomb. But they had had more excitement. There had been Japs there that they could see to strafe. And they had handed out cold justice, not letting anybody get away with it. Bruce had chased one Jap around a tree twice before getting him. He saw the Jap cutting for the woods and had let fly at him, chopping down half the tree. But that first time, the Jap had circled the tree fast enough.

So, Bruce, flying slowly and carefully, had come back on another pass and had straddled the tree, and when the Jap ran away from the bullets on one side, he ran head first into the bullets on the other side and was dead before he hit the ground. Hammond was waiting for the flights to return. He had the expression on his face of a mother hen whose chicks have taken to the water. When he saw Swede, he couldn't control his emotion.

"Are you all right?" he cried.

"No," Swede said, "I'm hungry."

"When I saw that oil," Hammond said, "I was so scared I didn't know what was going to happen."

"You were scared you would have to pay off all those bets."

"That's it." Hammond laughed. "That must have been why."

He laughed uproariously, all the nervousness going out of him in laughter.

There were sandwiches and coffee waiting. Nobody has time to cook on a ship during general quarters. Everybody has his battle station to man. The sandwiches were surprisingly bad that day. The theory was they had been purposely made bad to make sure they would last. This theory was voiced and there was chaffing about the bullet holes in the planes.

Each plane had some holes in it, and nobody could remember seeing any Jap guns shooting at him, so some of the fellows said, "Hell, we must have shot ourselves out of nervousness."

But Hansen (Ensign E. R. Hansen, the gunnery officer) disputed that.

"That's Jap stuff," he insisted expertly. "Our bullets make better holes than that."

"I wish I had seen them shooting," Ensign Aaron Katz said wistfully. "I'd have liked to blow their shooting back down."

And none of them said what they thought until Swede said it, saying, "Well, goddammit, we got some of them today, anyway."

And Ensign Robert E. Ries, even though he was from the Show-Me State of Missouri, agreed. "Yes," he said, "we made a start on it." Then Torpedo 8 took to the air again to look for more Japs. It was not yet eleven in the morning and the day was going to be long, as historic days always are.

Hundreds of square miles of Japs lay in holes yearning for darkness to put a cover over their wounds. How they must have ached for that day to end!

The Americans were not yearning at all, not for daylight or for darkness, but were just keeping busy doing whatever the situation required. Our fellows were very methodical about the death they were dealing out. They dispensed it in the firm, tranquil-seeming way of clerks behind a counter doing a job. This was the more remarkable, since under the outer-calm and under the outer-relentlessness there was a whole swarming jitter of nerves, veritable paroxyms of nerves, our fellows having been men a much longer time than they had been soldiers.

It was that way with Torpedo 8, too. As a squadron they were a special case in the struggle for Guadalcanal. They had a bill to present and collect and were determined to do it, but their determination had not transported them too far from the world in which the rest lived.

So the stubby, sausage-fat torpedo planes, as Swede led them for another try at Marmasike, were very businesslike-looking and the fellows in them looked very businesslike, but, underneath, their emotions rolled like the sea, and they themselves were crafts riding that sea, dipping into it and rising over it, putting their shoulders to it and throwing it behind them, or, where that failed, squashing the sea down.

In the beginning, over the carrier from which they took off, there was the same gray, heavy-laden weather curling soggily in the air like a soup simmering. It was hard to tell there that a war was going on: the air was so quiet and had that sleepy quality of a rainy day.

Then suddenly, as they neared Guadalcanal, the air became streaked and quivery with American voices: "There's trucks coming out of the woods there. Go down and strafe them!"

The question: "How many are burning?"

The answer: "Wait till I spit the glass out of my mouth. My window is broken up. There's a bullet stuck in it, and I can't see through it."

"Did you make 'em pay for that glass?"

"Well, I don't know. Wait till I go back. Yes, let's see now, there's a truck burning, I see three, four, seven trucks, burning, all burning, every one of them, frying."

"All right, go over the woods and see if you can spot any more hiding there, unless your broken window there ..."

"Oh, the broken window is no trouble. I'll just carry it with me. Most of it's in my lap, anyway."

The storm front was moving to south and west and it pressed down on Torpedo 8 and forced it down to about four hundred feet. Occasionally Jap voices spoke into the fellow's ears, making swift, blubbery sounds there, but they never saw the owners or the planes from which they spoke and just flew on placidly and firmly in a tight, wing-walk formation.

Near Hokiwai Island, they found a ship covered with palms. It was so well covered that, if the weather had not forced them down real low, they would never have seen it. First one saw it, then another, and pretty soon they were all talking about it, circling over it and trying to make out what it was.

"Well," said Swede, "there's only one thing sure about it; the Japs must want it really bad if they took all that trouble to hide it."

So Red Doggett flat-hatted over it and his bombardier, Lawrence, laid an egg right in the middle of it, and what it actually was nobody in Torpedo 8 will ever know now because it blew up and sank so fast.

After all the trouble to get there, the fellows found nothing in the way of a target at Marmasike and then high-tailed it over to Takataka Bay for a look-see. They were all learning fast. Their bombing was getting as good as it had been in practice and they were learning things about the country: the natives waved at them, the Japs didn't; where nobody in a village stayed to wave, that meant Japs; where a village was as prim and orderly as grass could be made by weaving, it meant natives; where it was sloppy and the clearings looked hasty, it meant Japs.

At Takataka Bay, Squadron 8's fliers found a sloppy, hasty clearing and nobody waving at them, so they went down and worked it over, doing it in style and at their leisure, plowing it up and blowing it up and digging holes for it to fall down into.

Then they started heading back for the carrier. There was, suddenly, a whole blurt of blubbery Jap voices in their ears, and one high-pitched squealing scream of a pilot hit and falling and sounding his pain into his radio, and an American voice saying, "There's a blue-plate special coming up on your port quarter. Get it!" and an American voice replying with nervous elation, "Raaawther!"

Torpedo 8 kept their eyes peeled for the Japs who seemed at that time to be all around them. But they couldn't see any and saw only people along the shoreline—a white shoreline jumbled up with jungle. The people were waving at them in the way that people wave at history going by, as if it were a circus parade. There were a few whites, missionaries, and those strange, generally surly and neurotic people known as South Sea traders, but most were black and had red frizzled hair and gleaming bone needles stuck through their noses.

Then Guadalcanal came into view. It was smoking. The Marines and the Navy down there sure looked as if they were cooking with gas. Long lines of landing boats were threading across the water to Tulagi on the north and Guadalcanal on the south. They looked from the air like ants marching. They had the same purposeful, ceaseless relentlessness.

At the time, as Torpedo 8 found out later, under the smoke over Tulagi a Jap was making a bid to kill with his native weapon: ju-jitsu. It was to be the first and last time, as far as any Intelligence officer on Guadalcanal could discover, that the Japs were to try their much-vaunted, much-dreaded ju-jitsu in hand-to-hand combat with our men on Guadalcanal. This Jap, a burly one, picked out a Marine stripling, an eighteen-year-old with the skinny bones and pale face of a city boy. He got the boy down and got the boy rolling, but the boy made the Jap roll with him. They rolled together down a twenty-five-foot incline.

Then the boy stood up. He was barehanded. He had lost all weapons except his nerve. But he stood up and the Jap

couldn't and never could get up at all after that. The photograph shows his head trampled unbelievably flat, so flat it is
impossible to believe a human head ever could get that
way.

But at the time Torpedo 8 knew nothing of these terrible
desperations and bustled briskly through the smoke and did a
careful hunt from Taivo Point to Mairu Sound. They saw two
anchored schooners, in good trim, but with no signs
of life on them. That meant Japs to Torpedo 8 by this
time, Japs hiding, so they hit both schooners, and one was
seen to be sinking by the time they pulled out of there.

When they got back to the carrier, they found the ship
bubbling over with mixed emotions. Pug Southerland of Scouting 5 had been shot down. Scouting 5 had gone in on Jap
bombers. Zeros had gone in on them.

There was a hot, crowded brawl and Pug had got hit in it
and had gone down in the middle of it. There was emotion over
that. Pug had been a good friend to all of Torpedo 8. (Inci-
dentally, Pug, at the time they were mourning him, was
making his way back, wounded, to Tulagi, confident that we
would have a good hold on that place by the time he got there.
And, as so many millions have found out about Americans,
his confidence was justified.)

There was emotion over the whole operation, the smoke of
it, the narrow escapes in it, the victories in it. It was a day of
great victory for everybody, even for the unskilled laborers of
war who had done nothing more than put their backs into
trundling a plane out to the tee.

The carrier that day had landed and put into the air more
planes than any ship in recorded history. The sandwiches
were still there and still, as Torpedo 8 recalls, bad. So, they
went to bed bubbling from them, too, from emotions and
sandwiches, and slept like separate logs.

First aid is administered to men wounded by grenades on Guadalcanal

5 - THE JAPS ATTACK

A TORPEDO PLANE CARRIES the most devastating weapon the war has yet devised against the ship, the torpedo, and is therefore the most valuable airplane of the sea. As such, it is expended carefully, even frugally when ships are not around, and so, since the Japs did not at first set up any ships for us, Torpedo 8 frittered restively and powerfully along the edges of the Guadalcanal landing operations.

They were like hungry sharks, prowling and nibbling along the shore. At six-fifteen in the rainy morning of August 8, they took off with Dave Shumway's Bombing 3 with orders to excavate the Japs from a hill on Tanambogu Island.

Tanambogu was the toughest nut the Marines found to crack during the occupation. The tiny island was almost filled by a single hill and the hill was almost filled with Japs holed up in impenetrable coral caves. The Japs in those caves taught the western world something we had not known before; that the Jap, man by man, is the toughest, least indomitable fighter in the world. He does not give up, but fights until killed.

When an explosion kills everybody around him and wounds and dazes him, he fights wounded and dazed until killed. He does not let his dead discourage him. He makes a wall of them to protect him.

He holds on to his mind, whatever the shocks against it, and bends it relentlessly to the single purpose of fighting. Even when he cannot hold on to his mind any longer, when the nerves in it have been shot out from under it, even then he does not surrender, he kills himself.

Our men learned this terrible truth bloodily in the caves in the hill on Tanambogu. They had to go into every cave separately and fight it out separately with each Jap and kill each Jap there separately. They threw dynamite in, and, where the cave was too deep for the dynamite to reach all the way through it, they ran into the smoke of the explosion with tommy-guns and bayonets and finished up that way.

Swede and Dave Shumway flew their fellows low over Tanambogu. They saw fires burning and lots of landing boats, some of them wrecked. A Kawanishi four-engined patrol bomber and Zeros lay heaped up among bomb craters and somberly smoking buildings.

Marines were waving at them, these, of course, being Marines who were not in a line of fire. Then Group Commander Thorne, whose office was a dive-bomber perched high in the air between Guadalcanal and Tulagi, told the pilots to wait a while, and, in the meantime, to do a submarine patrol. There were too many Marines embroiled on the hill and camped in cave mouths to make bombing of the hill anything but disastrous. So, for two hours, while Marines killed Japs and got killed, and pulled themselves slowly away from the target area, Torpedo 8 said, "Ho-hum, what a life!"—one actually said it, the rest thought it—and went growling through rain squalls up and down the coastline, looking for enemy subs.

The enemy subs were there, but they wouldn't come to the surface when airplanes were around, and that was the general idea of the patrol, to keep the subs down and harmless. At nine-thirty, Thorne whistled the boys back to work.

On the first pass, Thorne said, "Bring 'em up a little. You're falling short."

"Okay," said Swede, "watch this one," and went tearing for a cave mouth. But Andy Divine was just ahead of Swede and, as Swede points out, despite official record on the proceedings, it was Andy's bomb that dropped into that open mouth.

"That's fine work, boys," said Thorne; "you're really digging now."

In a few minutes, the whole hill looked shaken up, as if a big steam shovel had picked it up and put it down. The hill was still there, but it didn't look natural anymore. It looked as if it had been put together loosely.

However, bombs are not very good against a dug-in personnel that doesn't mind the noise bombs make nor the shaking-up they produce, and the Marines still had plenty of work to do when Torpedo 8, out of bombs, headed back for the carrier. There were eight-thousand-foot mountains to climb.

That took a lot of gas. And before the fellows got to the mountains, they knew they had no gas to spare, and that it would take plenty of luck for all the planes to get back safely. They had spent too much time making sure they'd do some good for the Marines. They had maneuvered too long and aimed too carefully.

In their anxiety to make sure, they had come in too low, too, and Swede's plane and Divine's plane had holes in them from fragments from their own bombs, and were flying heavily and awkwardly.

The Marines who had to stay there and finish up watched as the planes lifted themselves from that bucket of blood and wheeled into air that had been washed clean by rain and now was being shined to sparkling by the sun, and thought, "Lucky guys!"

The pilots hoped the Marines were right. Engines were leaned out all the way and the formation went along at the slowest possible speed. If Zeros had jumped them now, there wouldn't have been a prayer for any of them. If the carrier had been under attack, or had been chased by submarines, or had had to move on out for some other reason, or even if it was just having routine difficulties in landing planes, then goodbye Torpedo 8; goodbye to the planes, anyway, if not to the men. If there were thunderheads in the way that had to be got around, then that would be enough to dump them. Everything had to be just right.

Everything had to fall in exactly the best possible place, and that, Swede thought, was too much luck for a man to have. It wasn't too much to hope for, just too much to count on. Swede is not a religious man. He is like the rest of us and goes to church when he can't think of a reason not to. He sat there thinking of all the luck the squadron needed, and then thought that God had been very good to the squadron these last two days and had led it by the hand through all kinds of dangerous and narrow passages.

The sun was very bright now. The day was spread out like some sparkling, dappled, fathomlessly varied sea all over the sky, and Swede's thoughts ran quietly and darkly through it. The squadron, he thought, had a job of revenge to do. It had made only a start on that job. The rest of the work still lay ahead.

They had all had quite some nerve when, without consulting anything but their own feelings, they had taken on that job. Quite some arrogance there, thought Swede, for "Vengeance is mine, saith the Lord." That's what the Bible says.

Well, decided Swede, this would settle it. If God arranged everything to fall into the right place now, the carrier, all the planes using it, the Japs in their airplanes and submarines, the weather and the wind; if that happened, then Torpedo 8 hadn't been arrogant in its decision. Torpedo 8 had been prompted to make it, and had been right to make it.

All the planes landed all right. Swede had less than five minutes' gas left and one of the other planes coughed out, all empty, before it could finish taxiing. Then there were sandwiches, and then the whole air group was ordered out to spend the afternoon in practice around San Cristobal.

Bullet Lou Kirn's dive-bombers and Dave Shumway's fighters had been having thicker going than Torpedo 8, but they were ordered out, too. The football coach orders his squad to do six laps around the field after a hard scrimmage and the Navy orders its squadrons to practice even when the Bible tells them they are traveling on the beam.

Late in the afternoon of August 8, the carrier, its mission of attack accomplished, began moving out of the area. In the night of August 8-9, Jap warships moved down the channel, sneaked around Savo Island, surprised three of our cruisers—the Astoria, the Vincennes, and the Quincy—and the Australian cruiser Canberra and knocked them over with an avalanche of fire.

But the Japs had been surprised by us in the first place and were not yet ready to exploit this great gain they had made. Their task force killed our ships and killed our men and then withdrew to the Jap bases in the west and north, where a large sea and land force was being thrown together hastily to take Guadalcanal away from us and keep us from ever getting it back.

The Japs, eventually, were to throw against Guadalcanal forces greater than they used against Singapore and the whole Malay Peninsula. But this first force of theirs was only as large as the one which they, in their arrogance, had assigned to the task of taking Midway and, perhaps, Hawaii. It took the Japs two weeks to gather this force from its stations in the north, from as far away as the Philippines and China, and put it into position to move.

Torpedo 8 spent these two weeks in patrols, protecting the line of communications feeding our hastily established bases in the Guadalcanal area, and in the course of these two weeks the squadron suffered its first death. Ensign Cook's plane was the one the finger fell on. Cook was the baby of the squadron, the newly married one they all called "Cookie," and "Lookie, lookie, here comes Cookie." C. E. Thompson, a bombardier, died with him.

It was the wind that did it. The wind was in a querulous mood that morning. As Cook lunged down the deck of the carrier, his propeller scooped up wind and pressed it to the plane's sides.

A gust edged him too far off center and he found himself in a whirlpool of winds. The plane bobbed an instant helplessly on the whirling surface of the whirlpool while Cook fought forces too big for him.

Then it was over. The plane went nose first into the sea and kept on going straight down. The whole thing was over in an instant, before anybody even had time to let go a shout. There was a turret gunner who popped up suddenly while the tail of the plane was still sticking out of the water and began paddling and blowing. There was another fellow who had not belonged to the squadron, but had just gone along for the ride. He came up when the whole plane was all the way out of sight under the water.

Cook and Thompson didn't have that kind of luck. They kept riding on down to the bottom with the plane.

"Cookie was such a boyish man," said Swede in relating this incident, "that everybody just hated seeing him go, and Thompson had been a fellow everybody was for. That's what

Thompson had been, the kind everybody who knew him was for."

And in those two weeks Frenchy Fayle surprised himself and everybody else by winning a medal. Frenchy is the sort of boy who can't work when somebody is telling him to, and Swede had had some vexations over him in the course of the arduous practice sessions he had stepped Torpedo 8 through.

But now Frenchy, with nothing else to do but patrol and wait for the Japs to get strong enough to make a challenge, abruptly took it upon himself to solve a bomb shackle problem. Adapting the standard shackle to drop a bomb when in a certain position had been baffling the best mechanical minds on the carrier. It baffled Frenchy, too, but only for two days. Then he solved it. When Swede heard of this, he had a four-inch brass medal struck for Frenchy and presented it to him at a full-dress ceremony. Frenchy came to the ceremony not knowing what was going to happen next, but figuring sure he was going to catch it for something he had done or had forgotten to do, or maybe just on general principles.

Then he heard Swede's solemn voice intoning, "You are hereby awarded the "Medalle de Bomb Shackle" for meritorious ingenuity beyond the call of duty."

Everybody congratulated him with mock seriousness, but it is a fact that Frenchy's solution is now being used by the whole Navy. Abruptly, our carrier force changed its course and steamed north. The enemy had begun to move. As at Midway, the canny Jap moved with weather, hiding under dirty banks of it and converging with it on Guadalcanal. How much of him was there? More than eighty ships.

This was not a task force. This was a fleet. The firepower it contained staggers the imagination. To add it up in terms that can be understood by the mind that is land-based, and therefore has a limited take-off, is impossible. The nearest to it is to take the rough Army way of approximating the battle strength of a battleship to a full and fully armed division. And this was not one battleship coming down from the north and west. This was a whole army of ships. And how much of us was there? This is still information that is of value to the enemy.

But there was not nearly so much of us that we could ever hope to sink all, or even most, of the enemy's fleet. We could not afford the losses necessarily involved in such an attempt. If we had won, it might very well have been our last victory in the western Pacific. The best we could do was fight defensively—prevent the enemy from attaining his objective and punish him as much as possible for attempting to do so.

The enemy had the ships to sweep the sea clean and land on Guadalcanal. He was carrying with him the airplanes he needed to exploit a landing on Guadalcanal and make it good. This was an amphibious operation. The ships were no good to him once he'd got his beachhead on land. The planes were not much good to him until he'd got the beachhead. But, if we could strip him of his planes before he got his beachhead, then he'd turn his ships around and go away.

There would be no point in his expending ships to establish a beachhead that could not be exploited. It was the combination that he needed; ships to blow the door down and airplanes to go on through it, and if we knocked out any part of the combination, we'd win. He'd go home. We didn't have the ships to knock out the surface power in his combination. But, when you sink a carrier, you knock out all the planes that depend on it, and the Japs' carriers were vulnerable, as carriers, it seems, always will have to be in this war. Carriers have to be fast to stir up enough wind for take-offs.

They have to be airports; that is, they not only have to support planes, but support the enormous weight of men and supplies for servicing and, in part, maintaining them. There are limits to the amount of weight that can be pushed rapidly through the water and those limits are reached before a carrier can be armored and armed to stand off an attack like an ordinary warship.

So, air-power at sea inevitably has feet of clay, and it was at the feet that we

aimed in this battle—the second battle for the Solomons—as we had aimed at and hit them in our other great defensive victories of 1942, Midway and the Coral Sea. We did not destroy the Jap in those battles.

By stripping him of his carriers we simply made it impossible for him to realize on any sea-victories he might achieve, so he broke off his attempts to gain them. And that was enough for us in the very desperate days of 1942 when our Navy had had its back thrown up against the wall at the outset, and had to fight from there with fingers instead of fists. Late in the afternoon of August 23, a Sunday, a search plane radioed a report that Jap ships were only two hundred and seventy-five miles away from our force. Our planes got into the air at once, among them six from Torpedo 8 with Swede leading, but by the time they got to the designated area it was almost dark and no ships could be seen. The weather threw the planes around and hid the sea more often than not.

The planes searched until it was too late to go back to the carrier. Night landings on a carrier are not anything a man makes unless he has to, and this time, because we held Guadalcanal, the planes did not have to. They headed, instead, for Henderson Field, and became among the first American squadrons to use Guadalcanal for the purpose for which it was taken; as an unsinkable aircraft carrier anchored in the throat of Japan's South Pacific conquests.

The first time Swede saw what was to become 1942's most famous airport, Henderson Field, it looked to him like a strip of lumpy tape an inch wide and five inches long. Landing lights were sputtering along its edge. They consisted of Jap mineral water bottles filled with oil and rags that had been set on fire.

The earth of the airfield was soft and muddy and a brownish black. It had been a grassy plain before the Japs had leveled it off, and high, wiry grass still lay in patches between ground installations.

There was a mound on the edge of the field, about midway down its length, with a crazy-looking house set crazily on its top. Everybody called the house "The Pagoda," and, although the Japs shot at it a lot, and it sagged and tilted after every bombing and shelling, becoming more Pagodaed with each concussion, it was to remain there in all its shuddery bleakness until finally leveled the night of October 13.

The airfield ran pretty much from east to west and plunged headlong into jungle at either end. To the south, the soggy, densely overgrown earth heaved upward like an animal into mounds and ridges and hills, and finally into a full-bosomed stony mountain range eight thousand feet high. On the mountain range shacks, sluices, and shafts, that looked like a mine at work, could be picked out with glasses, and the story the men heard at once, and heard through all the days of fighting that followed, was that this was actually a gold mine run by Australians.

When the Japs had landed on Guadalcanal, the Australians and their native labor had been cut off from escape. But, while they couldn't get to the sea to escape because of the Japs, the Japs couldn't get to them without considerable difficulty because of the terrain. So, the Australians, without anything better to do, had simply gone on taking gold out of the mountain and piling it up. They were not sure for whom they were piling it up, and were not yet sure, even after the Americans had landed, so they just went on taking it out of the earth and piling it up, work being a very handy thing to have to do when trouble is around.

To the south of the airfield was a coconut grove, about a mile wide and ending on the shore of that nearly round basin of water between Guadalcanal and Tulagi which had already started earning its names of "Iron Bottom Bay" and "Sleepless Lagoon." At night, Guadalcanal, even at its most peaceful, looked ominous. There were so many men there and they were all so quiet. They lived in holes and in tents beside their guns, and they all seemed to be holding their breath, even when they were not.

There was a simple reason for this. Men habitually spoke in low tones on Guadalcanal, and even laughed in low tones, because everybody knew how important the sleep of others was, not only to the men sleeping, but to those around them. For on Guadalcanal, more than on any other place on earth at the time, it was crystal clear to the average American that his life depended not only on himself, but on the men around him. If sleeplessness clouded their reflexes, they would be dead, and he, too. But whatever the reason, there it was, a fearful battlefield, the earth of it gaped open by shells and bombs and burrowing men, thousands of them, thousands of shells, thousands of bombs and, thousands of men.

Occasionally, macaws squawked restlessly in the trees, or a cow lowed. Occasionally, trucks and jeeps crawled slowly across the earth, their headlights flicking on and off like eyes opening and shutting. Dim-out lights were not good enough in a terrain as treacherous with holes as that one was. Machines were too hard to come by to risk piling them up, even at the cost of giving the enemy a possible target. From the jungles to east and west of the field, other, indescribable sounds came.

Screams, for instance, wild, gurgling ones that split the whole night, and you could not tell whether it was some animal found by a larger one, or some American or some Jap; or whether it was just Japs trying, as they did in those days, to scare our men into giving their positions away. The men of Torpedo 8 found blankets and went to sleep on the ground under their planes. Bulldog Morrell of Jimmy Smith's fighter squadron was an old friend of Swede's and had tipped him off to that.

"The Japs are always around," he said. "This is a party. You don't have to go hunting for anything. It's all there on a plate. I keep my plane parked on the end of the field, right next to the tee. Because, let me tell you, when there's a scram take-off around here, scram is sure the word for it. Everybody knows that if you stay on the ground, you're not going to get a Jap. All you'll get staying on the ground is klunked."

Bulldog Morrell also suggested that the men hit the sack as rapidly as possible.

"It's quiet for a moment," he explained, "but you never know how long that is going to last, so the best thing is to get your sleep while the getting's good."

US Marines in the Guadalcanal Campaign

6 - The Second Battle for the Solomons

A GREAT MANY OF THE MEN of Torpedo 8, no doubt, felt that night on Henderson Field, Guadalcanal, that this could very well be their last on earth. And, at best, even if the best of luck was with them, they were in for a very remarkable time the next day. They were going violently to get the most violent kind of revenge for Midway. But the healthy human mind was a mechanism for escaping from what it can't do anything about anyway, except, maybe, worry. And this mechanism does its job skillfully and generally with success in the well-integrated kind of man the well-nourished American way has produced.

So, after getting chow, a very rugged chow ornamented in a slippery way by some captured Jap victuals, most of the fellows wandered a few hundred yards down to the Tenaru River where quite a big battle had been fought two days before and plenty of Jap souvenirs were available for the brisk bargainer.

The fellows got their souvenirs and then lay down on the ground under the planes. There was machine-gun and rifle fire going on sporadically all around them, and they argued awhile about which was ours and which theirs, and finally learned to tell the light-caliber Jap guns by the thick, plopping whistle they made as they went off. Then, just as they were drowsing off, a Jap submarine opened up on them with three-inch and five-inch shells and everybody made a scram take-off for a foxhole. The shells didn't land anywhere near them, and they spent some time arguing about just where those shells had landed, and about just how funny everybody had looked making tracks for a hole.

The argument kept getting quieter and quieter and pretty soon even the last man there was sound asleep, sleeping as deeply and sweetly on that awful earth as if he had not the least notion that the day coming might likely be his last.

In the morning, Torpedo 8 went back to the carrier which was maneuvering into a better position to hit the Japs who were still skulking under the banks of weather out there somewhere. Like so many of the great events of war, this vast, violent, venomous effort to strip the Jap battle fleet of its air-power began on a squeaky, petulant note. There was a doubt-ful contact report.

A search plane was radioing its base that it was seeing an enemy carrier force and our people had intercepted the mes-sage. It might have been spots before that unknown plane's eyes, but there the message was, and it had to be looked into.

Our people at sea could not look into it by remote control, or by asking anybody. If they used the radio, they'd give away their positions. They had to go see for themselves. Swede felt pretty sure it would be a wild-goose chase. He told Bruce Har-wood that, and told Bruce he had had enough of chasing the goose the day before. This would have to be Bruce's goose. Bruce took off after it with seven planes.

Then, about twenty minutes later, confirmation of the con-tact report came in. There were actually Jap ships out there, one of the task forces into which the battle fleet had split up on coming into the arena. But the task force was not in the place where Bruce was heading, and there was Bruce, taking seven of the deadliest known weapons against ships at sea for an airing over empty water.

Right on top of that came the news that opened up all the radios. The Japs were throwing a Sunday punch. The Sunday punch was not only on its way, but heading for the carrier. There was no point now in maintaining radio silence. The Japs knew where everything was. So, orders to cancel the strike were radioed to Bruce, and orders were given to all planes on board to get up into the air.

Bruce never got the message, fortunately, because if he had, the Jap battle fleet might never have been hit in the heart, or, anyway, would certainly not have been hit that day, and so, instead of turning back, the flight from Torpedo 8 went hustling on its way. But the planes on board got their or-ders all right and took off like a football from a boot, without waiting on the order of their going. And all this time Swede was in air plot, standing first on one foot and then the other, thinking one minute it would be maybe, and the next that it would be no, and never getting a yes, getting the wind, the weather, and the course, speed, bearing, and distance of the enemy force, but never getting a yes.

Until finally he got the yes and ran to the flight deck and found the deck uproarious with take-offs, the planes of Torpedo 8 in the melee and taking off, too.

"Clear the decks for action!"

That was the way it looked, like a movie illustration of that order, the gunners swinging out their guns, ripping the "bloomers" off them, and the plane handlers getting the planes the hell up into the air where they belonged. The plane crews weren't even waiting to get flight gear or navigation gear.

But Swede thought nix on that, and ran below for his stuff. This was what he had been priming himself for since June, and priming the whole squadron for, and he didn't intend to go off half-cocked on it, whatever the Japs were up to, and maybe flub up on the whole thing just because of some little something lacking, some little darn fool something like a pair of goggles to keep glare out of his eyes, or a pair of gloves to keep the reflexes in his fingers warm and smooth.

That had happened before, a whole attack flubbing up because of some little something a fellow had forgotten or hadn't thought important, but it wasn't going to happen to Swede. Not ever, if he could help it. Swede had quite a way to go for his gear and ran fast. He had always been a good runner. As we have mentioned, it was knowing he could run fast that had snarled up his football career at the Academy. As Swede came running back to the deck, Torpedo 8's last plane was starting to bunch itself for the run down the springboard. But Swede ran faster than he'd ever run before and got to the pilot before the pilot gave her the gun and pulled his rank on the man, thumbing that pilot out of there. It was Red Doggett.

Red climbed out grumpily and shouted something, but there was so much noise going on and so much uproar in Swede's mind that he couldn't figure out what Red had shouted until he got into the air and calmed down and pulled his thoughts together.

Then he figured out that Red, who was a chief and had to "sir" the commissioned officers whatever his feelings were at the time, had shouted,

"Aw, hell, sir. This ain't fair, sir. It's my plane, sir."

Up in the air, Swede counted noses. He saw that Taurman was flying his plane with his crew and Frenchy Fayle, Barnum, and Ries were all mixed up, too, but there. That was the important thing at the moment. They were there.

And they had been drilled hard enough so that they were all interchangeable. The particular machines they had to work with, and the particular men alongside them while they were working, did not matter so much, not when everybody knew his job, not when everybody knew everything there was to know about what he had to do.

"I hope," thought Swede, "I hope, I hope, I hope."

When Taurman got into the cockpit, he didn't look the ladies' man anymore. His girls would not have recognized him. He looked serious, and like a very deadly young American, and Frenchy didn't look fizzy, the way he did usually.

He looked all bottled up, and you'd never think, watching Barnum take out after Japs, that he could ever put his head back and howl, "Party, party, let's have a party."

And they were all especially unlike their natural selves now, because this thing that they were in was shaping up to be just like Midway. As at Midway, there was no fighter support. The fighter planes were being left to beat off the Jap assault on the American ships. And as Torpedo 8 said goodbye to them out of the corners of their eyes, and got a "good luck, good hunting," look from them in return, they all thought that that wouldn't be where they'd put the fighter planes if they had any say.

The way the Navy wants it, a lot of arithmetic goes into a torpedo bombing. There is the arithmetic of the attack, worked out in yards, feet, inches, and split seconds, and the arithmetic of the defense, worked out the same way, in equally split seconds. The primary objective of the torpedo plane is to get a torpedo into a ship. A torpedo not only hits harder than a two-thousand-pound bomb. It not only hits where the ship can least afford to take it, on the line Davy Jones pays off on.

But it also gathers the sea up to help it in its work of destruction.

A bomb cannot be aimed at the heart of a ship. It can get to it only with luck. But a torpedo is aimed for the heart and gets nowhere but into the heart, and bursts there with all the explosive and concussive force of a two-thousand-pound bomb, and all the fire of a two-thousand-pound bomb, and then it always has water rushing in on its heels to compound the injury done. It's a devil all right, a real, regular devil, working with fire and brimstone and floods.

So that's the major part of the job, to get the torpedo in. But a torpedo plane is expensive, and, as its crew gains battle experience, is not quickly replaced. No plane is ever better than its crew, and you can't turn out torpedo crews with a jig or on the home front. A man cannot learn what he has been taught until he gets into battle. That's where the secondary objective of the torpedo plane comes in.

The primary objective is to stay alive long enough to get the torpedo in. The secondary objective is to come away safely so that the Navy will have the use of you again. For all its bulk and terrible destructive power, power enough to take a ship by the back and break it like a dry bone, the aerial torpedo is cranky and temperamental. Once it is dropped from the plane, it is on its own. It runs under its own power with a propeller of its own, the propeller winding up the infernal machine and cocking it. Unless it's been dropped right, with an eye for waves, wind, height, and distance, the torpedo may kick up its heels on its solo run and work itself right out of the picture.

The torpedo pilot has to give that factor Number 1 priority in his mind. He has to come in very low over the water, to make the drop not too exciting for the torpedo, and has to keep his plane at just the right keel and has to drop at just the right moment. A target that maintains even altitude and even keel is the answer to an anti-aircraft gunner's prayer. Then the pilot has to reckon on the vexing fact that his torpedo is only a little faster than a warship.

The advantage in speed is actually no advantage at all because a torpedo cannot maneuver, and a warship can.

In fact, as airplanes have discovered, a warship under steam at sea can be as exasperating as mercury on the flat of the hand, while the best the torpedo will do, when its temperament has been salved by the proper handling in the factory, during shipment, from the plane, from shackles, from wind and from water, is run straight.

So, the pilot not only has to come in very low, but he has to go in real close, close enough practically to get his paint blistered by the heat of the gun's muzzles.

That way, the torpedo's run is short, and the warship has less time to use its talent for maneuvering. It has been reckoned that the normal reaction time for one of those vigorous young Navy men is one-tenth of a second.

That is, it takes about one-tenth of a second for him to do what he has to do after he sees what it is. The enemy, it's safe to say, has the same reaction time. He is a vigorous young man, too, and a great many of those Japs are quite some men. Anyway, when you're up against them, it's wiser to figure on their being quite some men. These petulant little fractions enter into the thing, too.

For, once the torpedo plane launches itself into battle, it's all reflex from then on. It's reflex on the plane, and on the ship the plane is trying to hit. There is no time for any other reaction. If a man's reflexes jam up in an action like that, it's the last thing they ever will do. The arithmetic of the reflex gives it all to the warship in dodging a torpedo, and, if that were all there was to it, there would be no point in sending a torpedo plane in.

But the torpedo pilot feints and maneuvers before he drops. He tries

to trick the ship into maneuvering into its own grave, and, since he cannot aim at where the ship is, but where he thinks it will be when his torpedo gets there, he has a fair chance of doing it. At least, he has a better chance than if the thing were decided entirely by the arithmetic of the reflex. However, outmaneuvering the torpedo is only the last shot in the warship's locker. It can do a number of other things before it is reduced to the desperation of having to get out of the way. Usually, it has fighter planes as its first line of defense.

Besides this, it invariably has a screen of other ships so placed as to be able to use the principles of converging fire to set up a wall of high explosives. They don't actually have to aim. They just set up the wall and let the torpedo plane run head-on into it. Anyway, that's the theory. The torpedo plane, coming in low, not only sets itself up for fighter planes to dive down on it, but sets itself up for every gun on the task force against it, including even the big twelve-inch and sixteen-inch rifles.

These rifles can fire into the sea and throw up splashes that can knock the plane down just as well as a shell itself. Besides this, torpedo planes, necessarily limited in number in the Pacific where there are only such limited facilities as carriers or jungle airports available for handling them, cannot diversify their attack. They concentrate on one ship in the force, which means that the force can devote all its firepower to erecting only the one wall of explosives, and does not have to thin its fire-power out over more than the one wall. As its answer to these assaults upon it by its target, the torpedo plane jinks and corkscrews. Jinking is a wabbling, flipping motion, and corkscrewing is exactly what the term implies; a twirling spiral, the plane advancing by going up and down in forward-moving circles. This makes the torpedo pilot a very busy man during his run. He knows that, if his plane remains on even keel for as little as twenty seconds, the enemy gunners will have as good a bead on him as if he were the side of a barn.

But he has to keep his plane on even keel to drop his torpedo. More than that, he knows the ship's officers will know he is preparing to drop when he gets on an even keel, and, when they see him assume an even keel, they will throw their ship out of his path.

So, assuming an even keel two or three or even more times during a run is part of the campaign by which the pilot hopes to flabbergast the target into maneuvering itself into its own grave.

And there the torpedo pilot sits, throwing his plane around with both hands and both feet, his eyes flitting from enemy plane to enemy ships to target to waves to altimeter to speedometer, his brain racing through the arithmetic of destruction and the arithmetic of conservation, and struggling with fractions and with the immense complications of the arithmetical laws of probabilities, and with all his reflexes working at once, thrusting past, around, and through each other, like the notes of some terrible symphony, an unwritten, unmemorizable, unlearnable one that he must conduct with the utmost precision and with utter flawlessness, knowing well that one flaw will kill him.

In the climax, when his massive task begins and he launches himself upon it, he finds himself confronted with a problem that makes all others look small in comparison. Like all soldiers in battle, he must suspend his imagination, or find his mind running riot and becoming incapable. But, unlike most men in battle, he must keep his mind busy with, and must think intensively about, exactly those things most calculated to galvanize his imagination.

He must figure out, for instance, "They are shooting at me there now, and will shoot at me here next, so I will go right where the last shot landed," and then stop his mind from figuring out the rest of the problem; what would happen if his figuring was false. To do this, to work the mind frantically and yet keep it in tight rein, to make it lunge from problem to problem with fierce recklessness but with iron-bound precision, cannot be taught by anyone or learned by anyone or acquired through luck or accident or hard work.

Doing it involves the whole sum of a man's personality which, in turn, is the sum of all that he is, and when a torpedo pilot goes in to deliver his torpedo, he is putting his whole self on the line and asking it to come through for him.

Some of the arithmetic in the situation is for him and some against him, some of the machinery is for him and some against him, but when the kind of person like himself is against him, that's the end of the story.

He is living his last moment. The flight of planes Swede was leading found a target at five minutes past six. It was very summery there.

The sea was darkening, but the light in the sky was low and the silhouettes of the enemy ships stuck up black against the light. The sunset there was tan and golden, and had big, fat, rolling curls of purple in it.

There was no carrier in the force. Swede would have preferred a carrier, but he was afraid to spend time looking. Night time does not come gently in the tropics.

The sunset is very brief. The sun falls into the sea and the night pounces like an animal and blinds everybody in it. This target had four heavy cruisers in it, plus six light cruisers, plus six destroyers. It wasn't anything to throw over the shoulder. It was quite a bird in the hand. In living right up with it, sort of climbing into a hammock with it and tousling with it there for five or six minutes, Torpedo 8 would be climbing into a hammock with a force that could throw out more tons of metal in those five or six minutes than a thousand planes could in a whole attack.

The four heavy cruisers steamed abeam in the center of the force. Three of the light cruisers were aft of them and three forward. The six destroyers champed and pawed around them in a rough, diamond-shaped formation, so that whichever way a plane came in there would be hoofs in its face. And the sleek, truculent ships steamed along blackly behind their guns and stuck up against the light sky like toys.

"Get the nearest big one," said Swede over the radio, and threw the throttle over as far as it would go, and pulled as much mercury as the motor would take. The plane bunched and sprang forward like a kicked horse.

"Get the nearest big one" were the only words uttered by anybody during the whole attack. Nobody wanted to take up any of his mind's time thinking of words.

And during the two months when Torpedo 8 had been waiting to get revenge for Midway, that was one of the things they had drilled on—to do without words.

They had expected it would be useful not to have to spend fractions of seconds thinking of words, and had so indoctrinated themselves with the team-work of attack that each man knew instantly, by what the flight leader was doing, just where to place himself and just what to do there. Torpedo 8 had been coming up from the southeast when the target was spotted.

The target was about fifteen miles off when first seen. The sky around the target was fish-belly white and clear of clouds, but there was a purple-colored curl of cloud northwest of the Jap task force. It looked as if it had been draped up there in the sky by somebody who wanted to get those Japs, for it was in exactly the position it ought to have been. Torpedo planes coming out of that cloud would have the cover of it during fifteen or twenty seconds of the most dangerous part of the run and then would come out of it on the nose to drop across the target's bow.

Swede saw the cloud and headed for it right away, with all he had, as if it were that Crescent Road cottage of his in Birmingham, and Taurman, Fayle, Barnum, and Ries tore right along with him, the five motors going like one with a steady thundering.

Swede saw little orange lights flickering at him from the ships, winking on and off like fireflies, and knew these were the guns firing at him, and waited a moment in his mind to see how close they were coming. It takes what seems a long time to both sides of the deal, to the fellows shooting the gun and the fellows being shot at by it, for anti-aircraft fire to climb up from muzzle to target.

Two dive-bombers had come up from the carrier with Torpedo 8, and Swede, while waiting in his mind for the shells to explode and show where they were, looked around for them and saw little black specks off a way and saw them get bigger until they got as big as small birds, and recognized them for Zeros coming toward him, two or three or maybe half a dozen Zeros, maybe more, off in the west there, tearing along to head him off.

Then strange black chrysanthemum-shaped puffs of smoke flowered in the sky directly ahead of him and began to walk toward him, each flower-like puff with large hurtling chunks of steel for petals, and he began to jink and corkscrew and thought, "Where the hell are those dive-bombers at," and figured they must be climbing, and wondered bitterly why it was that dive-bombers always felt they needed to dive off the top of the sky when eight thousand or ten thousand feet would be enough to give them all the velocity possible.

He couldn't hear anything except the motor working and he was so used to that he hardly heard it. The rubber-padded phones hung silently on his ears.

He kept forcing his throttle over as far as it could go, trying to push it past the farthest point, and kept one eye on the instrument panel and kept another on those ominous puffs of smoke. He hadn't the time to worry about the dive-bombers any more, just hope that they would be riding their fly ball down into the glove when the time came, and hadn't the time to worry about the Zeros coming up. That was his gunners' job, to worry about them, and he'd know soon enough that the Zeros were there when he heard the machine guns working. That was one sound he would surely hear, the soft, insistent chugging of the machine guns. Swede steered straight for the anti-aircraft fire. That would keep the Zeros off. They'd run around the anti-aircraft fire, he thought, and wait to pick him up when he came out of it.

This anti-aircraft fire was his funeral, he thought, why should they stick themselves into it, and next thing heard a whip-like crack slam viciously against his ears. The plane flew out from under his hands. The tail went up, the nose down, then sideways, then up and down at once, wiggling, as the concussion from a near burst tossed it in its irresistible currents. Swede fought the plane back into control and waited a single, endless moment to see how it flew, whether it was limping or not.

It flew with strange, whistling sighings from wind cutting itself in jagged shrapnel holes, but there was full-throated uproar in its going and it flew straight and powerfully, like an arrow.

"Okay," thought Swede, and got the plane jinking and corkscrewing again, and wrinkled his brow and thought, "Whew," and felt sweat squash in the wrinkles.

The cloud to hide behind seemed a long way off. It didn't seem to be getting any closer. Swede wondered, agonized, if the purple-colored air of the twilight had thrown off his depth perception, if they would not have had a better chance going right head-on into the target instead of swinging around and coming in astern of it. But that cloud had looked so good. Maybe that was what had happened.

The cloud had looked so good, he had wanted so much that it be in the right place, his hunger for that had thrown his judgment off and warped his depth perception, and maybe even made him put a cloud where there wasn't any at all.

They were in the middle of the anti-aircraft fire now. It was whipping and smacking all around them, ahead and under and over and behind them, and Swede worked on that. He threw his plane where the last shells had burst, figuring the gunners would correct their aim to where he had been, and did that a dozen times, each time being right, and then suddenly was in blank, utter quiet behind the cloud. He looked around a moment. The rest of the planes were right behind him, rubber-close in the attack formation they called ABC.

(It would be no service to Torpedo 8 at the present time to describe the attack formation they favor and use most often, but if you've ever seen a football backfield start moving for a lateral play where the idea is, if one is knocked over, another would be near him to take over the ball, then you will have the general idea of an ABC attack.)

The Zeros were out back somewhere, pouring on the coals and gaining.

Swede took all this in at a glance and took in the whole purpled-over peace of the air behind the cloud. Then he wheeled into a screaming power dive and hurtled into the clouds. The last thing he saw was the flight wheeling with him and, behind them, a Zero with its tracer bullets reaching out ahead of it like wilding, flinging fingers.

After that the cloud closed around him. The cloud flew up to the window in streams of scud at first, then thickened and pressed against the window and muffled the plane, then thinned again into scud that raced like birds, and finally, abruptly, was no more. They were out in the open now and had hit their approach right on the nose.

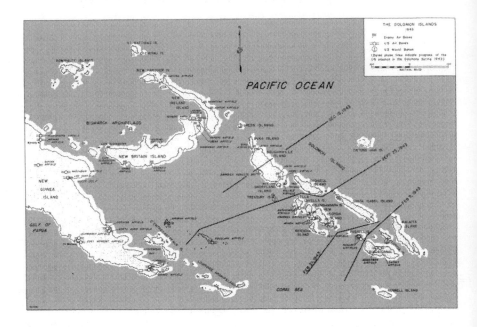

Map of Solomon Islands showing the Allied Advance in 1943

7 - Vengeance for Midway

THE JAP SHIPS WERE DEAD ahead of them. Swede thought of that for a moment. He had that moment. The Jap guns had been reaching for the planes in the cloud, but had been firing too high. The planes had gone up to the cloud at seven thousand feet and had been in the cloud about fifteen seconds and had come out of it skimming the water. Those planes were so low down now that a man could hang his hat on them. And the dive-bombers were not there. No, God knows where they were, but they were not there, and it was going to be like Midway, just like Midway with a rut to travel along and the Japs having nothing to do but pour enough steel into the rut to drown a plane, no less hit it.

The thought rose screaming in Swede, and he fought it down and looked around. The whole flight was right up there with him and Zeros, too. The Zeros were hot to get them and were riding the anti-aircraft with them, two, three, five Zeros streaking out of the clouds with scud streaming off their wings like the smoke of battle.

He had time to count five Zeros, or think he had counted five, when again the steadily jinking and corkscrewing plane flew out of his hands, its pedals butting roughly at his feet. The Jap gunners had dropped down on him and were bracketing him. For a minute, in his despair over the rut he was traveling down, he thought the hell with it, and went straight into the fire and saw the smoke of the explosion coil lazily over the windshield, and saw it torn away by the speed with which he was moving. Concussions were grabbing at his plane steadily now, as with hands, and his head jounced so that it was hard to see his instruments. But he had been shaken out of his despair.

"That's good," he thought of the concussions, "they make me harder to hit," and began diving toward bursts, first to one side, then to the other, and up and down, hoping that the Jap gunners would be smart enough to correct their aim, counting on them to be smart enough, knowing that if they were too dumb to do that they'd kill him sure.

The Jap ships were close now. Swede had a big black cruiser right up against his eyes. All the ships were keeping formation, waiting for torpedoes to drop and pumping their guns steadily. There must have been twenty thousand pairs of eyes there, straining to see what he was going to do and straining to act on what he was going to do. He leveled off and held level ten seconds. The ships wouldn't break formation. They didn't believe him. He leveled off again and held it and held it until his brain felt so full of fright it seemed bursting.

Still the Jap ships didn't break formation. Those Japs were tough, the toughest they could be, and fooling around wasn't going to get them. He was past the light cruisers now, in between them and some destroyers. They were throwing broadsides at him from both sides and stuff was going past him from both sides and killing Japs.

But they didn't care about that. They couldn't, not then, not with a torpedo hung up there in the belly of a plane for them to shoot at. That was the thing to get, the torpedo. Then the whole setup swam sluggishly into focus for Swede and he let go of his torpedo and swung away. He saw the Jap ships break formation and saw the little bubbling streak of the wake of another torpedo going through the water like a pencil stripe, and wishing it the best of luck, and threw himself into the job of getting away from there.

The Jap guns were still after him, chasing him, even over-running him, and he whipped and ducked and sashayed along like a little boy pumping his legs so fast they became bow-leg-ged. And right in the middle of that, a voice startled him, saying, "Nice work, Swede. You got him square in the belly," and he looked around open-mouthed before realizing the words had come through the earphones and the voice was Jack Elders', one of the dive-bombers.

Most of the anti-aircraft fire had stopped now. Only the heavy stuff was reaching this far. That meant the Zeros would be on him to chop him down any minute now. He looked around him. He couldn't see any Zeros. He couldn't see anything, not ships or any planes. The sky had darkened so rapidly in those five or six minutes of the attack that it was practically night now. It was very quiet in the sky. There was no feeling of death in it anymore. Whatever death there was going on far below on the surface of the sea did not come up into it.

"Is everybody all right?" said Swede into the interphone.

"Okay here, sir," the tunnel gunner said.

"Okay here, sir," said the turret gunner.

"That's good, because it was some piece of business down there and I didn't know... I mean, hell, well, I'm glad we're all right now... Does anybody see anything?"

There was no answer.

"Where is everybody?" cried Swede. "Where have they all gone to?"

"I saw some planes go down, sir," Lawrence, the tunnel gunner, said.

Swede was circling around slowly, staring into the swiftly gathering darkness.

"How many? Ours or theirs?"

"Two or three," Lawrence said. His voice seemed to be chattering. "Three, or maybe two, something like that. They hit the water and burned like a match was thrown in them. Maybe four of them, sir, or something like that."

"Were they Japs or us?"

"I don't know, I'm not sure," said Lawrence in his chattering voice.

Once the planes had dropped their torpedoes, the team broke up and it was every man for himself. The doors in the joint were all knocked down now. Each man chose his own exit and trampled the air down getting out through it.

Frenchy Fayle was going so fast, he thought if he ever turned, he'd break his plane in two. So, he threw himself right over the cruiser his torpedo was aiming at. His gunners worked a lather up on their machine-gun muzzles blowing down buckets of bullets at the ship. Then the fellows came out of it, they didn't know where, somewhere all alone in the middle of the quiet, empty evening.

Taurman couldn't get his bomb-bay doors up. They had opened all right and the torpedo had dropped through them all right, but then they wouldn't come up. Something had hit them and snarled them and Taurman had to go lumbering along like that through all that stuff, like a man trying to run with his pants fallen down around his ankles.

Then he saw what looked like Barnum and Ries going along ahead and above and far from him and he wallowed and lumbered in their direction. He knew Zeros were around somewhere, probably looking for more fight. Unless he was up in formation with the others, the Zeros would not get much fight out of him. They'd just be potting a pillow sailing clumsily in the wind. Swede made out the two dive-bombers high above the clouds, going off to the north where the last light of day still clung in a pale, tired way. He caught the final, blackly twinkling flicker of them as they dipped themselves into and twirled through some evil brew of their own.

A battleship, he heard over his radio, a Jap battleship that seemed to be trying to help out a burning Jap carrier. He knew now why the dive-bombers hadn't come in with him on the cruisers. They had been high enough up to see the other, fatter target and had wanted to grab it up, and had felt Swede's fellows could take care of themselves all right. And if not, if Swede's fellows couldn't, well, that was the young Navy's way of fighting—hit everything you can find and don't let anybody get away with it.

Swede understood their reasoning, but did not sympathize with it. His theory was to concentrate the attack and make sure of it. He knew that once the fellows got older and got to be old Navy and got over the excitement of all the plums the Japs held out for them to pluck, they'd stop trying to pluck them all at once and concentrate to make sure of getting one. Then Barnum and Ries came into place alongside him. He asked them if they were all right. They could hear him, but they couldn't answer. Their radios had been shot out.

They waggled and waved an "all okay," and the three planes circled some more, giving Fayle and Taurman another chance to join up. Fayle and Taurman kept groping around through the darkening night. They couldn't find anything. Swede kept calling them. They never heard him because their radios had gone, too. Swede was still calling them when machine guns started to chug.

There was a Jap biplane with twin floats hitting at them. Lawrence, the tunnel gunner, had his guns working well. Swede could see the tracers flying out into the night like cinders from an engine stack, hitting the Jap in pelts and ricocheting off him. That meant real slugs were going into him.

But he kept coming on. He was a very cheeky Jap. He made a rush low down and Swede saw him try to come up under and threw his tired mind into this new battle. He swung his plane in a tight turn.

The guns were silent a moment, then started chugging again, hesitantly, and then fell silent. The Jap was gone. He had either been shot down or had got swallowed up by the night, and was groping for them out there somewhere, blindly and stubbornly. The fellows kept looking for him, saying, "What's that," and "Is that him," and "Where'd he go," and then Swede said harshly, "Let's get the hell out of here." He was sick over Taurman and Fayle.

Fayle was adding up his gasoline. He knew he'd never get home now. He had no radio left and without a radio he couldn't find the carrier. He was trying to figure out how much more time he could give his luck to find him something to sit down on. Taurman had sent the tracers cindering up the sky and lumbered after them desperately. Then a lone Jap, on the prowl, jumped him and Taurman wheeled over and took bullets in the belly of his plane and squared off, like some dying bulldog, for the fight, but the Jap was gone.

He had pressed past his prey and had lost him somewhere in the night, and now he, too, must have been groping stubbornly. The whole night was like that, electric with a death that groped blindly for victims. And when Taurman got to where he had seen the tracers, everybody was gone, and he was alone again.

The exhausted men flew on in their separate ways, Swede with Barnum and Ries, Taurman and Fayle alone.

Then the moon came up, a big, fat, glistening moon, big enough and fat and glistering enough for a tropical cruise poster, and sky unrolled under it dimpled with chubby, silvered-over clouds. Swede's was the only radio that had kept working and now Lawrence said he couldn't make it go. It had been carrying on all right after the battle, but it just wouldn't work, and he didn't know what the matter with it was. He had tried to find out what was wrong, but he couldn't.

The interphone was going, and he told all this to Swede in the cracked, broken voice of a man who had been pushed past the limit and now felt there was no longer any use.

"There are nine men depending for their lives on that radio, including you," Swede pointed out. "I know, I know," cried Lawrence; "what do you think I've been telling myself all this time, sir?"

Taurman was the first to go into the water. He found the island of San Cristobal and went along it low to see if there were Japs on it. Nobody shot at the plane. "Well," he thought, "that means yes, or no, or maybe. Either there are no Japs or there are Japs and maybe they are afraid to shoot."

He passed back the word to the crew to guard their small arms and lowered the plane into the water so gently, so expertly, that nobody was hurt or even jarred very badly.

Then Frenchy went down. He found an island, too, a small one, and was afraid to leave it. He found it when he had only a few minutes' gasoline left and circled there, thinking, "Well, now, it will be easy; now all we have to do is live until we die." He was very reluctant to sit down.

A dead-stick landing in the water with a land plane is no job for an exhausted pilot. Still he was reluctant and clung to the air and clung and clung until at last he was very close to the water, and then suddenly was in it, scrambling for a raft that was bobbing alongside the drowning plane.

"My mother told me never to get my feet wet," the turret gunner remarked as he lifted himself into the raft.

"Where are the sharks?" thought Frenchy. He was very tired. He couldn't summon up the energy to get into the raft. The water was as warm as milk in a pail. He held onto the raft with one hand and drifted dully. "What's so easy about living until you die?" he thought. He looked around for sharks. "I'm sitting here up to my hips in sharks," he told himself, "and I don't even feel them. Where there's no sense there's no feeling."

"Mothers is the kind of people who are sore at the Navy for not issuing umbrellas," said the tunnel gunner as he began to ring out his skivvies. The moonlight came glancing across the water in a single shaft and seemed to look at the men like eyes wide open with innocent surprise. "It's not the radio that's gone bad," thought Swede. "It's nerves."

"It's a nice night for a swim," he said unexpectedly to Lawrence. "Did you ever go swimming in the moonlight?"

Lawrence stopped shaking over the radio and sat back helplessly. "No, sir," he said. Then he said, "Yes, sir, once."

Swede got him to tell what it was like. Ries and Barnum were in close. They could hear what was going on and they listened apathetically to Lawrence's story of swimming in a country river in darkness that made all the sedate elderly trees around it look like jungle. They couldn't make out what the whole thing was about, and the words just nagged at the tiredness in them.

At the end of the story, Swede said, "Boy, that must have been nice. You made it sound real nice."

"Well, it was nice," Lawrence replied. His voice wasn't chattering anymore.

"Look at that moon," cried Swede. "It looks like a mouth in the sky saying, "Oh," saying, "Oh boy." It looks like the "O," the great big, capital "O" in romance."

He laughed and Lawrence laughed, too.

"And the whole thing wasted," said Swede. "There's the moon. There's those beautiful little clouds to look at, like dimpled little baby pillows, the sea here and the tropics, and where's the girl? What a night! It makes you feel as if it had a waist and you could put your arm around it, and where's the waist! There isn't a waist within ten thousand miles of this night to put your arm around."

"Honolulu is only thirty-four hundred miles away, sir," said Lawrence and laughed again.

All the men in all the planes were silent a moment. The planes droned and grunted along, gobbling up the air under the moon.

"Try the radio now," suggested Swede. "Maybe the moon fixed it up. It's the only thing around here the moon could fix up."

The moon did it all right. With his nerves quieted, Lawrence had the radio fixed in no time. Later, while Taurman was going along San Cristobal to see if Japs would shoot at him and Fayle was bobbing dully in the milk-warm sea, Swede, Barnum, and Ries put their planes down safely on the carrier.

Bruce had a story to tell. He told it laconically. He had found the carrier Ryujo with some Zeros around it. The rest of this carrier's force of planes were off hitting our carrier. They had hit it, too, but not badly, not so badly that it would not be back in action again soon. The Ryujo had not been so fortunate. Torpedo 8 had left it smoking and burning. The Zeros hadn't been able to do a thing with them.

They had put one torpedo into it certainly and two more probably. The Army had hit it, too, and the Marines with bombs. Now, the Ryujo wasn't going to be any trouble after this. And everybody in the flight had got back safely; the planes shot up and wobbly, but safe.

"Was it tough?" asked Swede.

Bruce took a long time answering. He seemed to be searching for words at the bottom of his mind. "What do you think?" he replied at last.

"I think we got this thing licked," said Swede, "don't you?"

"It looks like it."

"If we can do this without any help, what are we going to do when we get help—dive-bombers, fighter support, coordinated attack? It will be a picnic. We'll go over them like ants over a picnic."

USS Yorktown at Pearl Harbor days before Midway

8 - TACKLING THE TOKYO EXPRESS

THE OFFICIAL WAR DIARY OF TORPEDO 8 discloses that on August 28 news was received that John Taurman and crew had been retrieved from their tropical retreat intact. All hands repaired promptly to the ice-cream bar of the carrier and had some on John. With, Swede recalls, hot fudge sauce.

So, the occasion was jubilant. Or at least it roared right along in the direction of jubilance until suddenly it was derailed. Somebody recollected that now nobody, but John would be able to enjoy the company of his girl. Her photograph showed her to be quite pretty.

There had been warm-hearted plans afoot among the squadron about her, all of them predicated on the assumption that John, being tucked away on some obscure tropical isle for the duration, would offer no competition. Now it was plain that, when the squadron got mainland leave, John would be again offering too much competition.

The War Diary then goes on to state that on August 30, Frenchy Fayle and crew were recovered. More roaring at the ice-cream bar. August 31, the Diary declares, the carrier got "pickled." This is the quaint Navy way of reporting that a submarine blew a hole into the carrier with a torpedo. For most people, a torpedoing would be an adventure remarkable enough to suffice quite a while. But the young men of Torpedo 8 were leading very crowded lives, so crowded there is hardly room left for the recollection of a mere torpedoing.

So, this adventure will have to be set down hazily. There was a jar. It came after general quarters had been secured. It felt like somebody had kicked the wall of your room hard enough to shake the whole room. The submarine had made a spread attack, firing five torpedoes in a broadside. Three had blown up in the water. One had just missed the ship and the fifth had hit aft on the starboard side.

There was, Swede recalls, "some running around." He remembers being annoyed by that. "Sit down!" he bellowed.

There was no place to sit, and he has a memory of men looking at him with round eyes and mouths rounded with surprise over his curious command.

"If you want to do something," he amended, "do something intelligent, like putting on your lifebelts."

However, lifebelts were not needed. The torpedo had had no luck. It had blown a considerable hole, but had killed no one and had not actually done anything a shipyard couldn't undo. After that, the carrier went home for patching and Torpedo 8 flew to a New Hebrides base. That was the beginning of their experience in using the islands of the South Seas for the purpose the times had intended them; as unsinkable aircraft carriers.

Two battles for the Solomons had now been fought. Three more remained to go before the Japs would finally admit that Guadalcanal was not for them. In the first battle, we had gained the island and lost four cruisers. Then the Japs had tried to do business on land and failed, and tried to do business on sea and failed. Now the Japs changed their tactics. Now they decided to concert their attacks and let loose a whole symphony of destruction, with planes as the wind instruments, ships as the brass section, and troops for percussion.

Through August, the Marines had been ranging over the island in a bush-fighting war. Japs would land in relatively small parties and parties of Marines would take out after them and wipe them out.

But toward the end of August and in September, the Japs put into operation their famous Tokyo Express, a sea train which ran nightly in the dark of the moon and landed troops in such quantities that small parties of Marines could no longer do anything against them.

Now it was no longer bush-fighting but war, engaging all the forces available. This posed a problem for us which was to the Japs' taste. Our available forces were required to protect Henderson Field. With the sea open to the Japs, we could not send our land forces chasing their land forces all over the island. We had to keep bunched up around the airport.

All the Japs had to do was throw a line around our bunch and build up their strength behind that line with the Tokyo

Express. This they did, and when they had accumulated what they felt was enough strength, they hit with it.

They stood on every platform there is for a fighting man to stand on nowadays—land, sea, and air—and threw their blows simultaneously, and our job was to kick the platform out from under them. Not from choice but from necessity, airplanes were given the job of derailing the Tokyo Express and so prevent the Japs from building up to battle strength. It was a job the airplane was not equal to, but, in those desperate days, we were very much the underdog in Pacific waters and simply could not spare the ships to do the job.

The Tokyo Express put out from way stations in Jap strongholds to the west and north of Guadalcanal. It came in the dark of the moon, wrapping its black shapes in the darkness. The Japs did not use transports. They used destroyers, frequently with a cruiser escort. They'd dump men and supplies on the beach a few miles up or down from Henderson Field, frequently so close that you could hear in those simmering tropical nights the soft putt-putt of the motors on their landing barges shuttling back and forth from ship to shore.

Then the warships would steam over to Henderson Field and "throw rocks," as our fellows used to describe it, trying to blow up the air force so that there wouldn't be anything looking for them when daylight came.

When daylight came, the Jap ships would be far away. The planes would go after them, but they seldom had any luck. There was too much weather to hide under, and too many tree-sheltered island coves to duck into. Henderson Field was so situated on Guadalcanal that ships heading for it, whether from the east or west, had to pass through a long relatively narrow channel between islands. This channel became known to Marine and Army fliers as "the slot" and to Navy fliers as "the groove."

During September and a good part of October and November, almost all the hazardous moments in our fliers' sometimes short, sometimes long, seeming lifetimes were spent in "the groove" mauling at the Tokyo Express.

It was, while it lasted, an extraordinary business alto-gether, the waters, the sky, and the land being a-prowl with sudden death during the "lulls" between battles, and being aflame with death during battles. During the "lull" periods no-body could ever be sure what was going to hit him, or when or where, and many an American and many more Japs stumbled unexpectedly into their graves. During the battle periods, eve-rybody knew what was going to hit him, and when and where.

But that didn't make it any easier to take. Torpedo 8 set up shop right in the middle of this business and set to work getting all the revenge they could.

Torpedo 8 went hustling up the groove the morning of September 16. It was a very bright morning with a feel of mid-day languor already in its air at eight-thirty.

Some of the fellows, as they walked to work down the slope from the ready tent over the soft, black, chuffed-up earth of Guadalcanal, couldn't help thinking of football going on back home, and of practice in the warm afternoons of Septem-ber, and of how soft the earth felt and steamy and aromatic.

Barnum taxied right into a shell hole at the start, so he was out of the attack, cracked right out of it. There were plenty of shell holes and bomb craters all around Guadalcanal by then. The planes ran lightly among them and ran past the amputated, mangled carcasses of wrecked planes left where they had fallen—some Jap planes, but mostly American—and lumbered heavily in the sausage-like way of torpedo planes into the crisp crackle of blue from the sky.

Then the planes circled. As the tunnel gunners got ready, guns stuck out and waggled like thin feelers from the plump bellies. The planes circled over Bloody Ridge and over the Tenaru River. Here tanks, artillery, mortars, and machine guns had been heaving at each other for days now, right on the edge of Henderson Field, so close to the edge that if you taxied too far on a landing you had the feeling, you'd run smack into Jap guns.

The Marines lay on their backs and looked at them and the Japs threw angry, futile sputters of machine-gun and rifle fire at them. Our artillery was working with the loud whopping whooms of artillery, and the shells went one at a time right through or under or over the planes. This always seemed something in the nature of a miracle to me when I was there. Planes, going to work, were always crossing in the path of shells going to work elsewhere; Jap shells and our shells, Jap planes and our planes.

And there the planes and shells were, rocking along through the same air to different destinations, but never did they bother each other. There were no traffic lights. The traffic was all-which-way in six different ways—north, south, east, west, and up and down—and Jap shells didn't mind, really, if by accident they clipped one of our planes and our shells didn't mind if they clipped one of their planes. But that never happened.

The planes and the shells went only where they were going. Torpedo 8 ground slowly up to a fairish amount of altitude and then lit out up the groove, going like a bee who had suddenly found out where the honey was. There were dive-bombers along, and this time it figured to be the picnic Torpedo 8 had been waiting for. Not really a picnic, just the regulation line of work.

But a regulation job of work, the kind of job a man expected to get and was trained to, always seemed a picnic in those wild early days in the Pacific where everybody had to do what had to be done, not what he was supposed to do. And when anybody got to do what he was supposed to do, he had that picnic feeling, or that riding-on-plush feeling.

The honey our fellows were looking for was made up of three Jap light cruisers and four destroyers. The feeling among veterans of combat is that the second time is the worst. The first time you don't know what to expect and always imagine worse than you get, or, if you don't, you're too keyed up to pay attention, anyway. But the second time, you know just what's ahead of you, you've got the cold dope on that, and carry into action, not only the facts of the situation, but the memory of your previous shocks. By the third action, you're beginning to get used to the business you're in. However, it didn't work out that way with Torpedo 8.

They were all raspy from lack of sleep. Nobody slept well on Guadalcanal. The Jap land forces there were night workers, making their attacks at night, and we were day workers, punching the clock at dawn and keeping on punching until dusk. Neither side ever considered changing its habits to allow for sleep.

And, still further complicating the sleeping arrangements, was the fact that the Japs were sure to bomb us somewhere between nine in the morning and four in the afternoon, and sure to shell us from the sea between eleven at night and four the following morning.

Besides being raspy from lack of sleep, the fellows had the trouble of a great glitter of sunlight on the blue water below, making ships as difficult to spot as pins in a tray of diamonds. But they did have that picnicky feeling from Bullet Lou Kirn's dive-bombers being along, so, at least there was not much trouble along the nerves about this being the second torpedo attack.

The target was located at nine-forty-five in the morning, up there somewhere between Bougainville and Choiseul. The three cruisers were in a column and the destroyers were splayed out skittishly around them. There were good clouds for Torpedo 8 to hide behind, but Swede decided to play it cagey and not hide at all, and just go in slow and casual in a friendly way.

The torpedo planes held back. The dive-bombers hit first. With the cruisers steaming in line, Swede maneuvered to come in on them on the bow. That way they could turn only to port or starboard to evade the torpedoes and either way they turned; they would be lined up broadside for the attack.

The squadron came in straight. They did not jink or cork-screw. They hung in the sky like the side of a barn for any Jap gunners who cared to fire. But the ruse was working. The Japs concentrated on the dive-bombers and decided the torpedo planes were friendly, or, anyway, would surely go into a cloud before launching their attack. It was very nervous going along like that, wide-open for an attack, about to become a dead pigeon any time any trigger on any one of the hundreds of guns down there was pulled. But Swede stuck to it and the fellows stuck with him, floating along, when WHAM! the trigger was pulled and the whole sky boiled up and thrashed and erupted under anti-aircraft fire.

Then Torpedo 8 raced. They were only three thousand yards away from the ships by then. That was what had given them away; the fact that they had come so close and were still coming.

But at that distance there were only fifteen seconds to go before the torpedoes could be launched and the Japs knew it. The Japs are lightning calculators, too, as good as we are at the arithmetic of death, and seemed to know that, even if they killed off every plane of the squadron, it was too late now to stop the torpedoes from dropping. They had waited too long. They had been tricked too well and the arithmetic seemed to unnerve them. Their guns hit nothing.

The gunners were concentrating on thinking, "Too late, too late," and thinking it no doubt with that high, dragged-out, wailing feeling that comes to all men when they know they're going to get it, and can't do anything about it but wait those seconds for it to come. And the Jap cruiser command seemed all unnerved, too. For the three ships in line wheeled a stately line to starboard, clinging in their turn to that same foolish, fatal line. The vision of the last cruiser was obscured by those ahead of it, and that was the one that got all its insides torn out and was left gaped open and floundering awkwardly in the water like some disemboweled fish.

The fire was very heavy on the way out. This was personal. This was carrying on the vendetta begun at Midway. The Japs ignored all the dive-bombers and really went gunning for the torpedo planes. This is something that happens all the time. The enemy doesn't like any part of us, but he seems now to like torpedo planes least of all. We've killed off more of his torpedo planes than any other type in proportion to number, and, no doubt, he is eager to do the same.

But it is also something personal that seems to exist in all Jap sailors, for our torpedo planes have killed off proportionately more ships than any other type plane. So, on the way out there was plenty of heat and one of Kirn's dive-bombing boys got all snarled up in it, pulling out of his dive alongside Swede and just a little ahead of him.

Swede had a good view of the boy's tail. It was whumping right along, and, just behind it, the anti-aircraft fire was bursting, so close to Swede he could see the streaks of blood-colored red in the black smoke. The bursts seem to give the dive-bomber's tail extra whump, and altogether that boy was like a dog legging it along so fast he makes his backbone wiggly. Swede had to laugh. He laughed very hard until he suddenly realized that if this were happening to the dive-bomber it must be happening to him, too. Then he stopped laughing.

But by that time, they were out of the fire and in cool, tranquil air. All the planes were out of the fire, and they all had a good time riding the breeze back home, and thinking what a picnic life is when there's enough stuff on hand to do an attack up right.

9 - FIGHTING THE WEATHER

THE JAPS KEPT COMING at night and kept trying to hide by day. Our planes would go all over everywhere looking for them, and, when somebody would spot a Jap force, the aviators had a job of work on their hands and ran to it. More often than not, they'd run, and get nowhere. What the search plane had seen would turn out to be islands, or the ships would just vamoose off somewhere, or weather would come up and block out everything there was below it or in it.

The ships kept coming through the weather. In the morning, up by Kokumbona or Cape Esperance (the Jap-held parts of Guadalcanal beach) there was the spoor of the ships glittering festively in the early light; brand-new metal landing boats for taking men and supplies off vessels and dumping them into the laps of the Marines. Then the planes would take out again, knowing Jap ships must be within minutes (airplane time) of Henderson Field, and sometimes they'd find what they were looking for and fetch it a clout over the ears, but more often they wouldn't.

That was the life there as lived in the air. It wasn't anything for a man to get fat on. You never knew, when you took off, what you were going to run into. The sky and the water were alive with enemies, and whether you stumbled into them or not, the pull on the nerves was there, that jittery, fluttery tugging going on in you that the aviators call "butterflies in the stomach." It went on when you were on the way out and went on when you were on the way in.

There was always a temptation to relax on the way in, unbutton your nerves and sit back in the cockpit because here was home and nothing left to do but land on it. However, the Japs fixed that last little moment, too. A couple of times they dropped down out of clouds on fellows who were just greasing in for a landing and that was an easy shot for them. That was just chopping meat in a bowl.

So, nobody ever relaxed until the motor was dead and he was walking back up the slope to Operations where two ounces of brandy was waiting, or, if that had given out, a bottle of warm beer. And even then, it took persistence to really let go of yourself. For there was always shooting going on at Guadalcanal or bombing or shelling in those days, and anywhere from a third to one-half the time the stuff was being aimed in your direction.

Torpedo 8 lived in tents they had put up themselves in a coconut grove. There they learned some things about war that a man can't learn flying around in the air looking down at the subject. The coconut grove was nicely pruned of brush and creepers so that you could count on seeing snipers before they got to where they could hurt you.

But bombs falling there, and shells, had a bad habit of hitting the trees and bursting in the air overhead and showering their stuff in a downward spray instead of upward as when they hit the ground. That made the holes you lay in small protection. Fellows lying in holes out in open ground had to have a direct hit to get killed, but when you were under trees you had a good chance of getting it from shrapnel.

Then the men acquired a more accurate appreciation of bombs, and of the difficulties in the way of doing damage with them. An aviator who has never been bombed or strafed from the air is likely to think of himself as cutting wide swaths into enemy flesh every time he lets go with his guns or bomb release. But Torpedo 8 found out different. They found out how astonishingly little damage bombs can do, particularly to people. The guns we had on Guadalcanal in those days didn't have any of the latest gadgets on them for blind shooting, but, during bombings, they generally managed to kill more Japs in the air than the Japs did of us on the ground.

In the meantime, between bombings and shellings and searches for targets that weren't there and hits at targets found by accident, Swede got dengue fever and lay cooking in it out of the picture, while Bruce Harwood took over.

On October 1, Engel, Dye, Divine, and Ries had a go at four destroyers who were legging it up the groove. The arrangement was as before. There were dive-bombers along to divide the fire, and, as before, the torpedo planes got in very close before the Japs decided they were not friendly planes, and opened up on them. Ries hit the leading destroyer a fully-loaded smack, and, on the way out, Divine's plane, which had been walloped pretty hard, waddled off the southern shore of Guadalcanal, and he lived a relaxing jungle life for a few days until found.

Two days later, Ries again, and Evarts and Mears, found a heavy cruiser and two destroyers in the groove, northwest of Russell Island, and went in on it. This was at five-thirty-five in the afternoon, and the cruiser was left smoking and listing in the water.

The three planes got out of there all right, and then Mears swung around and went back to have a look at what the cruiser was doing. That, it turned out, was a mistake. A Jap float biplane was in the area, and everybody was so busy looking at the cruiser that they didn't see the Jap plane until bullets were smacking all around them. The Jap made a mistake, too. He was too hot for the kill. He pressed his advantage of surprise too far, and Hicks, the turret gunner, got a cone of bullets into him and the Jap plane just blew apart in the air.

They all saw it go, a little black figure of a man popping up and out into the air like a pea shucked from a pod, and the plane simply bursting open and falling in pieces. That is, they all saw it except Deitsch, the tunnel gunner. The Jap had got a twenty-millimeter shell into Deitsch's head and he was lying unconscious over his gun. Deitsch, by some miracle, didn't die, but that day was the last day of war he will know.

The day after, October 4, Harwood, Esders, Taurman, and Doggett went after the same cruiser. The first attack had crippled it, and it hadn't been able to get out of the area. There was a very low ceiling and the fellows had a job of going along a rut again, but they did it. The wounded cruiser managed to flounder out of the way of the first torpedo, but it took the second and third right in the middle, while the fourth torpedo is listed as a 'probable.' That is, the last seen of it, it was about to hit. But nobody actually saw it hit because everybody was too busy hauling themselves away from there.

October 5, Harwood took a flight up to Rekata Bay to drop some bombs there. The weather was the worst so far. Only

Harwood, Mears, and Evart were able to get through it, and then Mears' bombs would not release. The fellows did some good, though. Hicks got himself another Jap float plane, and Struble, a tunnel gunner, shot one down, too. Actually, when you added it up, weather was the worst of it. It was always there, and when, by some odd oversight, it wasn't, it sure could come up in a hurry.

The weather blinded the men in the sky. They, of course, had instruments to 'see' with, but flying on instruments is not a joy in enemy country and over islands never mapped before the war and mapped during the war only in hit-or-miss patches as war missions carried planes over these areas. Particularly is it not a joy where you have to go low, sometimes skimming the water, to give yourself a chance to see ships and keep out of those muscular thunderheads.

And when you go low, you find out that black clouds there have a habit of sitting right on the water, and mist has a habit of gathering around the feet of an island and covering those feet so that the island seems to be a black cloud suspended in gray air. And altogether, it's hard to be sure, sometimes until it's too late, whether you're up against fluff that can't hurt you or a volcano or mountain that is going to kill you.

The nights were even tougher. There were nights that were so thick a fellow taking off into them was flying on instruments before he got to the end of the runway. Finding your way home on such a night was a job to make anybody gray.

For, about fifteen seconds or so, plane time, south of Henderson Field were eight thousand feet of mountains, and a man groping around low, looking for the field, say with his instruments thrown all out of whack from the jarrings of machine-gun fire, knew he had only those fifteen seconds to fool around with. If he banked too soon, he'd never find anything, and if he banked too late, he'd knock his brains out.

The fellows tried to get their bearings by picking out the thin line on the shore where the water ended, and land began. When the wind was kicking up the water into a sort of mild, baby-bodied little surf, this line of demarcation was a kind of pencil stripe, tiny and frail and wabbling and sometimes fading out altogether.

But, usually, there was no fuss-up in the waters of that lake-like lagoon and the only difference between land and water from the air was that the water

was a different shade of black. The land was a black black and the water was an off black, a kind of midnight blue.

Women can get the idea of what a strain this business imposed on the men if they remember trying to match up fabrics in very bad light. Only, they must add to their exasperations the fact that their lives depend on being right, and the fact that they have only seconds to make their decisions and act on them.

The fellows all had top-grade, twenty-twenty eyes and could make out the difference when they were low enough. The trouble with getting low enough was the trees. The trees were one height in one place and another height in another place. It was impossible to tell beforehand just which place you were coming in at.

Then suddenly the trees showed up in front of you, kind of as if someone had penciled in some lines on a blot of black ink. Trees can kill a plane just as dead as shells or bullets or mountains. It's happened time and again that a fellow with black all around him, going deeper into the black to look for the difference between midnight blue and black, finds suddenly these sinister little grayed-over black pencil streaks dead ahead of him.

There's this black blankness fore and aft, black blankness topside and below, black blankness to port and starboard, with just the littlest little bits of differences in the shades of blankness, such little bits of differences that you can't tell whether they're really there or were just caused by using your eyes too much. This business of getting spots and streaks and so forth in front of your eyes; any man who has been around planes in the Guadalcanal area any length of time has a real horror of them.

They give him the creeps. There was a Flying Fortress once, coming in for a landing after hitting a Jap carrier. It was a rainy night, the weather just socked in, as they say, and everybody was on the field sweating the fellows in, which is what the air force calls standing around with your heart in your mouth, praying everything will turn out all right.

This plane, nine men in it, came in perfectly, right in the groove, and was sitting down nice and easy right in the middle of the groove, when suddenly the pilot threw it to the right. The plane slewed around as if it had hit a stone when all it had hit was a nervous spot in front of the pilot's eyes.

A great scream of "NO! NO!" went up from all of us. But the pilot, of course, couldn't hear and smashed right into the trees on the right of the airfield and only four men were lifted out of the wreckage alive.

Well, that's what the fellows had to do when they came in from hunting Japs on those moonless nights that the Japs chose for their activity.

When a fellow threw his plane—fast, frantic fast because there's not a second to lose at the speed with which planes travel—away from one black blankness because it seemed stocked up with substance, he had to throw it into another black blankness and trust to luck or the angels that there was nothing in that one.

And all this searching and probing and matching up of different shades of black and worrying and acting with God for a guide must be about what happened to John Taurman and the two fellows with him on the last night they ever spent in the air. The night of October 6, about ten o'clock, Taurman went up to Cape Esperance with Esders and Doggett to bomb some newly landed Jap supply dumps on the beach.

There was a theory around that a fellow who had gone through the harrowing of a water landing rated a home leave or, at least, a leave in Australia or Hawaii. But the Japs were shaping up for a big battle and no man could be spared, so Taurman and his crew had been put right back to work. He had grumbled a little about it.

"Are you unfit for combat duty?" Swede asked him. "If you are, say so and I'll see what I can do about getting you a leave."

John had grinned back amiably. "Well, I can't go so far as to say I'm unfit," he had replied. "I just feel leave would be a handy thing to have."

"Who doesn't?"

"That's right," John had agreed.

On this Tuesday night, the weather was really socked in. Cape Esperance was only a minute or so flying time from Henderson Field, but it was easier to see the land o' Goshen from there than the airport. Dive-bombers went along to drop flares on the target. Dive-bombing on a night such as that one was is like roller-coasting blindfolded into the mouth of a volcano. But the flares were dropped where wanted and the torpedo bombers came in right on top of them. They did glide-bombing. Esders got a direct hit, a regular boom-boomer followed by fires.

"I think I'm a little high," Red Doggett said. "I'm going down to have a look."

His landing lights dipped into the blackness and then dipped more steeply. Esders banked around and suddenly he saw a reflection of Red's exhaust stacks bouncing and bobbing on the darkness and streaking along it. Red couldn't see the reflections. They were back of him.

It took Esders a moment to realize that the reflections were coming off the water and that Red, still going down to see where he was, was going right into the water.

"Red!" he screamed into the radio. "Red, for Christ sake!"

And then Red's plane hit the water a smack and bounced a hundred and fifty feet into the air and turned over on its back before dropping into the sea again.

Lawrence and Hayes were in the plane with Red. Esders sat still, holding himself in, waiting for Red's plane to blow up or catch on fire. But nothing like that happened. The darkness below remained dark. Then Esders began to circle, going lower and lower, looking for Red.

Taurman had been with him at the start. Then Taurman was gone, groping off by himself somewhere in the darkness, and Esders circled alone before finally coming back in. The weather curled in thick, gaseous, noxious-smelling rolls over the damp earth of Henderson Field.

Bruce Harwood thought maybe Taurman would come back with the dive-bombers and he stood in the weather, sweating the dive-bombers in. You could hear the planes in the sky, fumbling blindly one by one.

They threw on the searchlights, but twenty feet off the ground, the lights hit like up against a brick wall. The lights were walked uncertainly along the underside of the overcast, looking for holes to throw a beam up into. But the overcast was solid. The searchlights got nowhere, and the planes never saw them until their wheels were stretching out for the ground. Listening to the planes groping around overhead was an agony. The weather did funny things to the sound. It bounced it along and made it echo as in caverns and, if you were trying to place a plane by its sound, you wound up sure it was right in the mountains.

Finally, all the planes were accounted for, except John Taurman's. He was still wandering around up there. Bruce said, the hell with sweating, he'd go out and bring him back. Some of Lou Kirn's dive-bombing boys said to count them in on that and they all took off together, going on instruments before they got to the end of the runway.

The planes all kept together at first and about twenty minutes out ran into a crystal-clear patch of air with, right in the middle of it, a Jap destroyer that opened up on them with all its guns at once. Kirn's fellows had bombs with them. Bruce's plane wasn't loaded. He had just taken the first Avenger that was at hand, not figuring to run into anything. So Kirn's men grouped themselves for an attack and Bruce went on by himself.

As Bruce swung out of range of the gunfire, the dive-bombers were beginning to drop on the target. He saw the first bomb hit and flare like a match against the destroyer's deck. Then the weather swallowed him up. It was very eerie there, like being in the middle of a cave filled with cotton. He kept talking to John through the radio and then, abruptly, John answered him.

"Bruce," he said, "where in the hell are you?"

"I'm here," said Bruce. He felt so happy at finding John that, for a moment, he couldn't make sense. "Where are you?"

"I'm here, too, but I don't know where here is at."

The planes must have been very close together, but they couldn't see each other. Bruce gave him a course. "Just follow me in," he said. "I'll take you home all right."

"I don't know whether I'm following you or going away from you or going into you or what."

"Just keep talking, that's all. There's clear weather about ten minutes from here with a dead Jap in the middle of it."

"How do you know he's dead?"

"Dead or dying. Kirn's boys are hitting him."

"If he's dying, he's got company. That's the way I feel now."

"Don't talk like that," said Bruce. "But keep talking." He knew that, as long as they could hear each other, they were in the same neighborhood and not going away from each other.

"I've had so many narrow escapes tonight going down low to look where I am that I think I've used up all my luck," said John.

Those were the last words heard from John. Lieutenant John Taurman, R. J. Bradley, A.R.M. Third Class, and J. Robak, seaman First Class, went into the water finally somewhere in the vicinity of San Cristobal. Their plane had run out of gasoline. They were all pretty well dead-beat by then, but John had iron in him and fought the plane to a perfect landing in the trough of a wave and the three got out all right and got into a rubber boat.

Sometime later in the night, they heard a plane overhead. It might have been Bruce, or it might have been a Jap.

The fellows shot off a flare, anyway. Something went wrong with it and instead of shooting into the air, it shot right through the bottom of the boat, tearing the bottom wide open. The rim of the boat kept afloat, and the three young men sat on the rim, their feet in the water. They had no food and no water, and nothing at all to substitute or keep the sun off or make life consist of anything but sitting up straight with feet in the water. Everything they had taken along had gone down with the bottom of the boat and that crazy flare. If they wanted to sleep, they had to sleep in the water with one arm clutching the raft.

This kept on for two days. The sun was very hot in the daytime and the wind was chilly raw at night. They could see land far off every once in a while. That was the worst of it. The current there carried them in a big ellipse, swinging them along the shore and close to it, then out to sea, then back to shore. After two days of it, Bradley said he was going to swim.

"At least," he declared, "you have a chance that way. This way, we'll be riding around on a merry-go-round until we're skeletons. Then our skeletons will be riding around on the merry-go-round."

By that time Taurman and Robak were too weak to argue with him. Bradley went over the side. "Keep your chins up," he said. "I'll send help and a case of ice-cold beer." Bradley swam across the current.

For a long time, he couldn't get very far. He used up the best of his strength and still was close enough to the boat to hear the queer sucking noises the water made as it splashed against it. But he kept on. He figured he might just as well die fast in the water as take the long way around by staying in the boat.

Then the boat kept getting farther and farther away and Bradley thought he had won. He was gaining ground, he thought. But the shore didn't come any closer, and finally Bradley figured it out, that the boat had come to the end of the ellipse and was swinging around the corner of it now and going out to sea. Bradley wasn't gaining, but at least he wasn't losing.

There were plenty of fish in the water, strangely shaped creatures with colors like a pretty parasol. After a while, Bradley was alone with them in the silent sea. He swam until he was tired. Then he floated. After floating a little bit, he started swimming again until he was tired again. He was afraid to float very long, because every time he floated the current got hold of him and he lost ground. But he got tired more and more easily and had to float longer and longer.

It was hard to keep from getting desperate. He knew if he got desperate, he'd sprint and then all his strength would go out that way and he'd be helpless. He swam steadily. His arms seemed to weigh tons and, when he kicked his legs, the pain of doing so traveled all the way up into his head, and there was real pain in moving his arms.

After eight hours of that, he fainted, from pain and weakness and drifted like a corpse in the sea. Then a miracle happened to him. He had got past the ellipse into a new kind of current and this current carried him along over the coral reefs and washed him slowly onto the shore. Natives found him there.

His body was a mass of pain. Where it had not been burned into blisters by the saltwater, it was cut by coral. The natives were very puzzled that he was still alive. When Bradley told about the boat caught in the ellipse, the natives put out after it in canoes. But they couldn't find it, and nobody ever did find the boat again or Taurman or Robak. Swede got the news the next day and climbed out of bed.

"I want to go back up there, sir," he told Admiral Fitch, Commander of Air in the South Pacific, "and put some planes together and do some damage."

He and Earnest flew back to Guadalcanal together, six hundred miles by dead reckoning because they couldn't wait for a navigator. Swede always flew with Earnest, the veteran of Midway, when he could.

Earnest was rock steady in a formation. They hit Guadalcanal on the nose. It seems to have been quite a thing to do. Two more planes tried it the next day and got lost and had to sit down in the water. After that, Admiral Fitch stopped that kind of business, rush or no rush.

Admiral Aubrey Watch Fitch (1883-1978)

10 - FAST AND FURIOUS

ON OCTOBER 8, SWEDE, HARWOOD, Evarts, and Earnest bombed Japs on Guadalcanal and then hurried back to Henderson Field to be fitted with torpedoes. A Jap heavy cruiser and six or seven destroyers were coming down the groove, with plenty of Jap planes buzzing over them. The Marines had taken a beautiful Samurai sword from a dead Jap officer the day before and Major-General Vandergrift came over to the airfield with it and put it into the Operations tent. "The man who hits that cruiser gets that sword," he said.

Larsen, Earnest, Mears, and Katz went up the groove on that mission. Roy Simpler's (Lieutenant-Commander Roy C. Simpler) Navy fighters were along and so were Jimmy Smith's Marine fighters. The Japs were playing it smart that day. The cruiser was in the middle and the destroyers formed an inverted U around it.

It was like a tonsil in the throat of a gaped-open mouth and anybody going after the tonsil had to run the gauntlet of a very toothy mouth on the way in and on the way out. Overhead were the Jap planes—float Zeros and float biplanes—and the Marines dove into them while Simpler's boys came down low and strafed the warships to take their gunners' minds off the torpedo bombers. Then Torpedo 8 zoomed into the Jap mouth and split up there and came in on two sides of the cruiser. The cruiser, prodded into the jitters, made a complete circle and came around in its own wake like a dog chasing its tail. Then, as it put its bow into the wake and churned the water there whiter, it took a torpedo and stopped so suddenly it seemed to shudder in place. The destroyers lining the Jap mouth were laying down a brick fence of bullets, shooting into each other, but—which was the important thing to them—sealing up the mouth with the torpedo planes in it.

But Torpedo 8 didn't lose its head. It didn't go blind with excitement. It didn't try to go out the way it had come in. Instead, it lifted itself up through the roof with a chandelle, climbing straight up and flipping over and away while the

Japs kept shooting steadily at each other, waiting for the planes to run into their fire.

Nobody was sure whether Swede's torpedo or Earnest's was the one that had hit. So, Earnest took the cash (the sword) and Swede got stuck with the credit (a mention in the communique.)

And so, home. But not to bed. The Japs kept making home life on Guadalcanal very peppery with bombs and shells. Every time there was a daytime bombing, our planes staged a scramble takeoff. That is, everybody ran for his plane as fast as he could and then made his plane run as fast as it could, and if there was a traffic cop in the way, out there trying to regulate traffic... well, that was his hard luck. To those fellows, scramble meant scram.

Then, October 10, there were more Japs in the groove—two light cruisers and four destroyers. The fellows took off after them at dawn, their heads all light and droning from lack of sleep. There was a good, very hot sun that day.

Swede kept the fellows out of it. There is a popular notion that the sun is a good place to launch an attack out of, because it blinds the gunners. But it is also a risky place. Unless your plane is hung up on a straight line between the sun and the gunner's eyes, he is going to catch the glitter on your plane's wings. The glitter can be seen from as much as twenty-five miles off and is enough to give any gunner his mark. A gunner does not shoot at a plane. He shoots at the direction in which the plane is traveling. They found the Japs at seven-twenty-five in the morning, a hundred and seventy-five miles from Lunga Point. There were a few clouds in the sky, but they were too high up for the torpedo planes to use. Larsen, Earnest, Barnum, Hansen, Ries, and Evarts were on that run. The ships were in two columns, one cruiser and two destroyers to each column.

They picked out the lead ship in the port column. Swede used two commands.

"Form column," he said first. Then he said, "Close up." After that, he tore for the target.

The dive-bombers had dropped their fifth bomb by that time and burning planes were beginning to fall. One of them fluttered by Swede just as he was in the upward twirl of a corkscrew maneuver. It came so close that he could see the pilot with his forehead resting peacefully on the instrument board—dead, probably, but looking like a man sleeping. The sight threw off Swede's arithmetic.

He had had only a brief flash and he couldn't tell whether it was a Jap or one of his friends. (The odds are that it was a Jap because eight of their planes were shot down in those few minutes of attack, by our ten fighters while we lost only one.) But that odd and tragic picture of the man resting peacefully through the last moment before a crash into the sea disintegrated him, clung in Swede's mind tenaciously through the whole assault, no matter how he tried to throw it off.

The fellows don't think they hit anything that day. They were too tired. Their reflexes were too slow, and the arithmetic of death got numb in them and would not move for them. They let the business go as "a probable hit on stern of light cruiser."

The next day, October 11, the bloodiest, most desperate month in the history of our Navy began. During the month running from October 11, 1942, to November 15, the Japanese Navy suffered the greatest catastrophe ever inflicted on any great navy up to that time.

They were the attacking force, having by far the superior strength. They formed their own pattern for destruction. The month broke down into three separate actions, each connected with the other and dependent on the other, but each marked by a short breathing spell for regrouping and re-maneuvering.

On October 11, the Japs elaborated a pattern of attack which they were to duplicate again on November 11. From October 22 to October 27, they fought a battle, the type of which they never again would have a chance to duplicate on Guadalcanal. The battle beginning on October 11, known as the third battle for the Solomons, was primarily a sea and air action designed to land an overwhelming invasion force on Guadalcanal as reinforcements for the detachments already landed by the Tokyo Express. The tactics were simple. The Jap sent warships on ahead of the main force to bombard Guadalcanal and knock out its air-power.

Then, that done, their troops could land at will and their ships could be unloaded at leisure. The Japs scored a partial success in that battle. Their first attempt to bombard Guadalcanal, the night of October 11-12, was met off Savo Island by units of our fleet. This was the action in which the cruisers Salt Lake City and Helena distinguished themselves so greatly and the Boise suffered such cruel punishment.

The Japs left six of their ships at the bottom of Iron-Bottom Bay that night and ran away, and our task force withdrew. But the Japs came back again the next night and that was where they scored their partial success.

Our Navy was not there to meet them. Air-power had to do the job alone, and the bulk of our air-power never got out from under the Jap naval guns. Tojo managed to land a part of his invasion force.

Evidently, he believed that he had landed in overwhelming strength, for in the action beginning October 22 and enduring through the early hours of October 27, he hit with it. That was the fourth battle for the Solomons in which we lost the carrier Hornet and a destroyer, while the Japs suffered damage to two of their carriers and to other warships which our Navy refrains cautiously from claiming as sunk.

The days between the third and fourth battles were spent by the Japs in deploying their forces and were spent by us in feeling them out and bombing and shelling them. In the battle itself, the Japs fought with tanks, artillery, bayoneting infantry, and planes on land, and with warships and planes on sea.

We fought them off. The Japs made a breakthrough to the airfield on land and stripped us of our air-power at sea. But then they had to withdraw.

Their strength had been too depleted, as the military phrase is, for sweeping up all the dead into one bundle. So, the Japs had to try again. They went back to their original October 11 pattern, and on November 11 moved again for the fifth and climactic battle for the Solomons. This time our Navy met them in every play they made, and the result was total disaster; twenty-eight ships sunk (more than the Germans and the British combined lost at Jutland), anywhere from forty thousand to sixty thousand dead.

We lost seven destroyers and two cruisers, but the Japs gave up on Guadalcanal after that. They are a people who, for political reasons, value face more than lives, but they had no face left after that, and had to admit defeat to their people openly.

Torpedo 8 figured in all these passionate and terrible battles. Late in the night of October 11, naval gunfire heralded the beginning of the third battle of the Solomons. It was a coal-black night. The muzzle flashes streaked it, sporadically at first, then steadily as Admiral Norman Scott's force found the range of the enemy and found his soft meat. The fight was off Savo Island.

That night, the Marines started calling Iron-Bottom Bay "Sleepless Lagoon." Distance muted the ferocious crack of the guns into something soft and rolling like kettledrums. Once in a while, a lurid orange light blew up into the muzzle flashes and hung in it like a Chinese lantern in a lightning storm. That was a ship exploding. The guns awakened the fellows and they got out of their tents and sat on the edges of dugouts and watched the play of the battle in the sky.

"We ought to go down to the beach," Katz said. And Earnest said, that was right, they ought to, there was a hundred-million-dollar drama going on down there and they could have front-row seats.

But nobody went down to the beach. They were all too tired to move. They wanted to save their energy for battle. After about twenty-five minutes, the muzzle flashes began to dwindle off.

Then, five minutes later, they stopped and there was silence. The night came down sweet and calm over the arena. The fellows sat a little while in the silence, waiting to see what would happen. Hansen sat with head bowed over his chest and eyes closed. His whole face seemed sunken in fatigue.

"That's a hell of a thing to do to a hundred-million-dollar show," said Lieutenant Grosscup, "to be too lazy to open your eyes to look at it."

Then the fellows all went back to bed.

"A lot of guys killed out there just now," said Swede. He lay flat on his cot. He couldn't yet get out of his mind the peaceful figure of the falling aviator.

"Our guys or theirs?" asked Katz.

"Both, I guess."

"More of theirs, maybe. If it'd been more of ours, we'd have been hearing from their ships by now. They'd have been pasting us here."

"That's right," said Swede. He closed his eyes, but even with his eyes closed he could see the peacefully falling aviator.

The next morning, there was a dawn take-off. Fighters, dive-bombers, and Torpedo 8. They passed over water in which debris was floating. There were miles and miles of wreckage and oil streaks big and broad as highways. The oil streaks were the blood poured out the night before by killed ships and by wounded ships staggering away from battle.

From the air, they looked like roads running over the sea, sometimes straight, sometimes with crazy careenings and loops. Then the battlefield look of the sea ended and the groove assumed its ordinary aspect; peaceful-seeming, sunlit, jungle-lined and drowsy as any back road of the world. About a hundred and fifty miles up, they saw their target—two heavy cruisers with a destroyer escort. The cruisers had split-Y stacks and the light was such that the Y-pipes on the stacks could be made out clearly from a long way off. Again, the dive-bombers went in first.

There were seventeen of them this time, and Swede took his time about making his run and was very careful about finding a spot from which to launch it. He couldn't shake the tired, fretful, raw-nerved feeling out of himself. He wound around the sky for a long time, circling the inferno blowing up from the assaulted and ferociously assaulting ships. There wasn't much choice of springboard.

The sky was wide-open and empty, and finally, Swede thought, "The hell with it; this is as good as any," and turned around and went in from where he was.

Suddenly he felt he was going to get killed. He had never had that feeling before. He had been scared, but he had never thought it out plainly, that this time he was going to get it. Now the thought came to him unbidden and stuck in his mind as he lowered himself into the fire. The thought raced along into the fire with him, jinking and corkscrewing as he did.

The anti-aircraft wasn't any worse than usual. But he just thought it and couldn't stop thinking it, and the thought wailed through the arithmetic going on in his mind, and screamed through it and shook and bit and nagged at him.

Then he saw that the cruiser he was aiming at had made a mistake. Fooled into thinking the torpedoes were going to drop the first time the planes held level, it committed itself to a port turn. It amounted to turning its belly to a bayonet that had not yet been thrust.

"I'm going to take you with me when I go," thought Swede and launched his torpedo. The torpedo shot out in front of him and, suicidally, Swede did not turn away from it, but kept going right along with it.

The torpedo went along a few feet and then suddenly rose to the surface and turned its tail on the cruiser. It was either a defective torpedo or Swede hadn't dropped it properly.

Swede turned around then to get out of there. The feeling of death was greater in him than ever. He hated to turn around. He had wanted to take it head-on rather than get it in the back. As he turned, he saw two more torpedoes pass astern of that luck-laden cruiser. Then the cruiser's luck ended. Hansen's torpedo dug deep into it aft of amidships and its heart broke with a huge white spewing.

"Got it! Got it!" thought Swede.

He was feeling a little crazy by that time and he threw his plane around crazily. The plane looked, one of the other fellows said, as if it were doing an adagio dance with the anti-aircraft shells. Then Swede realized the anti-aircraft fire had stopped, and he was not dead. Katz was alongside him. Down below, he could see the cruiser, dead in the water and listing heavily and smoking like some dying fire.

"Sink, you bastard!" Swede screamed into his radio. "Sink!"

The ship just kept listing heavily. He could make out Japs running on it, some dragging hoses and some carrying the wounded and dead.

"Why don't you lay down and die!" screamed Swede. Katz had a twenty-millimeter shell in the thrust section of his engine. "It should make me lose all my oil," he told Swede. But he stayed, too. He thought he might just as well go down where he was as on the way home. That way he, too, could see the cruiser sink.

"Come on, Swede," Kirn said. Lou had gathered up his dive-bombers and they were circling with him. "You got your hit. Let's go home. There are plenty Japs around here."

"I know," Swede said. "There are a million Japs and they're all listening in on this radio. Come on, you Jap bastards! Come on and get us."

"I'm not losing any oil," announced Katz gleefully. Then he, too, began to scream over the radio at the Japs. "Bring your fat-tailed Admiral!" cried Swede. "Bring Tojo. Bring Hirohito and his fish guts. Come on, you bastards, come on you Yamomotos, come and get us." Katz and Swede kept this up for twenty minutes. Then they went home.

The cruiser still remained afloat, listing and smoking and motionless in the water. The crazy feeling didn't go out of Swede. He wanted that cruiser. He wanted it on the bottom more than he wanted to live. But anti-aircraft fire had damaged all Torpedo 8's planes except his own.

So, Swede, when he heard that dive-bombers and fighters were going up to make another attack, went along by himself. It was a wild thing to do, but he figured that the fighters would strafe the ships and help him out that way.

Lieutenant Jansen, leading Fighter Five on this attack, had more important things to worry about than a lone torpedo plane. He had the dive-bombers to protect from Zeros. He climbed up high with them.

"Jens," cried Swede into the radio, "come down and strafe for me, please."

But the fighter planes kept grumbling and grinding their way upward.

Swede had launched himself into his run now. "Jens! Jens!" he implored, "come down and get those gunners off me. I'm all by myself here."

Jens didn't even answer. There were Zeros swarming and stinging among the dive-bombers, trying to get them before they went into their drop.

"Jens! Jens!" Swede shook his microphone up and down with exasperation. Suddenly, by accident, the mike tipped the torpedo release and the torpedo dropped awkwardly, then lunged on its way. Swede was eighteen hundred yards away from his target. He had not even been aiming. The torpedo passed harmlessly astern of the cruiser. Swede was so angry he almost wept. Curses came out of him and the tears stayed in, while the dive-bombers dropped one by one on the dead cruiser's escort vessels.

"The cruiser!" begged Swede. "Get the goddam cruiser!"

With intense misery, Swede realized that war is not the time for doing a friend a favor. The dive-bombers weren't wasting bombs on the cruiser. They were being practical and aiming for the unhit ships.

By the time Swede reached home, he had exhausted the craziness in himself and felt all collapsed. "One more day like this," he thought, "and I'll be a real old man."

11 - Airmen Can Fight from the Ground

DURING THE NIGHT OF TUESDAY, October 13, starting at about eleven o'clock, a Jap battleship force stood off Lunga Beach and bombarded Henderson Field.

They threw high explosives in diameters ranging from five to fourteen inches for more than three hours. Then they steamed jubilantly away.

There were men who, to ease their nerves by giving themselves something to do besides crouch and quake, tried to count the number of shells fired per minute. But the explosions were so intertwined their task was impossible. General Vandergrift's estimate is "something between fifty thousand and one hundred thousand."

The Japs, working rapidly, methodically, and with the precision of trained, unhurried, and calm gunners, exploded these shells in an area about two miles long and a mile and a half or so wide. They walked their shells up and down this area and back and forth across it until there wasn't a single blade of grass that had not felt at least the hot breath of an explosion.

It was the most concentrated bombardment Americans had endured in this war up to that time, and, according to veterans of both wars, matched anything the Germans visited upon the Americans in France.

That is unarguable to anyone who was there. It is possible to imagine a bombardment like it, but impossible to imagine one that exceeded it. Each shell of the bombardment traveled, through the ears, as shells have a habit of doing, like a drill through a naked nerve. The salvos roared like a hundred express trains going abreast at full throttle through a rock tunnel. Their noise overwhelmed all emotion and all sense of pain, and left in a man nothing but terror and the even more desperate feeling of fighting to control terror.

The firing was continuous for twenty minutes. Then there was a lull of about a minute and a half while the Jap ships turned to steam back along the area. During this lull,

Henderson Field came awake with the figures of running, crouching men looking for a better hole. After that, there came another twenty minutes of continuously exploding uproar.

There were about twenty seconds between the time of the muzzle flashes and the time of the explosion. The muzzle flashes flickered, vast as heat lightning up into the sky. And before the glare of it subsided in your eyes, the earth under you began to tremble. It trembled more and more, and your ears filled up with sound. Then your head filled up and filled to bursting with sound, with a huge, whistling, rushing roar that grew greater rapidly with whooping shrieks. Then the explosion hit.

The ground shook so much under you that it threw you off itself like a massage machine, and you didn't know sometimes, when the explosion was close and the flesh of your brain had been flattened out with sound, whether you were just being shaken off the ground or being blown up onto the road to Kingdom.

After that, came the concussions. They ran howling and flopping over the earth like dying animals and in their midst was the slither of shrapnel. Millions and millions of pieces of shrapnel, weighing as much as twenty-two pounds and turning the earth like plows.

Torpedo 8's camp was in a coconut grove. That, they discovered, was the worst place to be during a bombardment. Shells hit the treetops and burst downward instead of, as when they hit the ground, bursting upward. Staying in the foxholes under the trees was about as much help as going into the open to keep out of the rain. During lulls, the fellows worked their way out of the area toward the beach.

Finally, most of them got into a big shell-hole and stayed there. It wasn't smart. One shell would have wiped out a good part of the squadron. But the noise didn't give anyone's brains a chance to be smart, or do anything but flatten out. These fellows had been in a lot of trouble together. They wanted to be together in this, too.

Swede was wondering how many planes would be fit for flying in the morning. There were fires going in patches all over Henderson Field and in the clear, lavender-colored light of Jap star shells smoke could be seen piling up into the sky. One of the men in the shell-hole was crying. He couldn't help himself. It wasn't like crying at all, but like a drunk with a crying jag. There was the same steady helplessness, and in the light of the star shells, there could be seen on his face a look as if part of him were ashamed for crying.

"Well," thought Swede, "anyway, we made them sorry they pulled the trigger at Midway."

Some Marines had come back a little while before with a report on territory they occupied immediately after Torpedo 8 had bombed it. They had had to bury four hundred and seven Japs there. And there had been dead Japs on every ship they had hit. Then there was the strafing they had done and the land bombing. Nobody would ever know the score Torpedo 8 had made, but it was respectable. That was sure; very respectable. Hammond was talking to the boy with the crying jag. He talked about the new song going around—"Say a Prayer for a Pal on Guadalcanal"—and about the best way to make coffee. Boil the grounds right into the pot. Then, when it comes to a boil, settle the grounds with eggshells.

"If you had pearls to swap for eggshells," he said, "you couldn't make the swap here, because there's no eggs, and where there's no eggs, there's hardly ever any shells either."

So, Hammond was using cold water, he said. A little douse of cold water and the grounds went shivering downward like a guy under a cold shower. The boy kept on crying and Hammond kept on talking. Hammond had grown a big, black crinkly beard and his words came softly and soothingly through the smother of his beard.

"If you'd shave off that beard," said one of the men, "you could be a mother to me."

The boy cried for more than three hours after the bombardment was over, and, while the rest of Torpedo 8 slept the sleep of nervous exhaustion, Hammond stayed up with him talking soothingly through the smother of his beard until at last the crying stopped and the boy fell asleep. When the bombardment was finished, thousands of Army men, Marines, and Navy men crawled out of their holes in the shelled area. They

were green-faced and damp with the sweat of agony, but unhurt. Only eight of them had been killed and eighteen
wounded. The rest were anxious to kill any Jap they could get
their hands on.

This seems to be true, that, if you are lying in a hole, a
shell or a bomb has to fall right on you to kill you. If it falls on
you, it does not do damage to anything else around you, which
might not be any comfort to you, but is, in fact, discouraging
to your enemy.

That is how, in the end, when tanks, airplanes, and artillery have done their best, the infantry remains the Queen of
Battle. Somebody has to come in on his two legs and stamp
around. However, in the morning, Torpedo 8 discovered it had
no planes left. Some had suffered direct hits, and, of these,
only holes could be found to mark the places where they had
once been.

Others had been minced up into a hash by shrapnel, and
Peterkin and Hammond, after careful investigation, decided
that with the best of luck only three torpedo planes would ever
fly again. It looked like Bataan all over again. Some of the fellows said it and more of them thought it. The Japs were
landing brazenly up the beach within sight of our positions
and they had a line out to keep our ground forces from getting
at them.

Major Jack Cram of the Marines made a suicide torpedo
run in a PBY on two transports. The PBY, a dignified, portly,
and leisurely plane, jinked like an old lady in corsets trying to
do a Highland Fling.

But we were using everything we had that day and the
next, including unarmed guts, and Major Cram got away with
it. He delivered his torpedoes and came home. He still doesn't
know how or why he got away with it. A dive-bomber went rat-
tling and groaning up to the tee, then paused. A major came
running to find out what was the matter.

"My hydraulics and hand crank both have been shot up,"
the pilot said. "I can't get the landing gear up or the flaps."

The Major had to scream to make himself heard above the
motor. "You see those Japs there," he pointed over the hill to

the beach. "They're coming here to kill you and me. It's a question of killing them first."

"I know," the pilot said. "I wasn't complaining. I was just telling you."

Then the dive-bomber took off. With the landing flaps down, he couldn't get up any speed.

His plane went a few yards down the runway and then swooped up into the air like a badminton bird. Everybody ran to the hill to see how he made out against the transports. He dropped on them like a feather. He couldn't dive at all, not power-dive with his landing gear and landing flaps sticking out into the air and holding him back.

But he came down vertically just the same, twisting slowly and gently like a falling feather, and he dropped square and came home, too, without knowing

just how or why. Swede thought it looked like Bataan, too. He had orders to evacuate all the Torpedo 8 crews he could not use.

"Hell," he said, "we don't need planes to get Japs. The Japs are coming to us now. We don't have to fly after them. All we have to do is stay where we are and shoot them down."

Some of the fellows left on the destroyer MacFarland. The Japs dive-bombed it before it started to move and knocked off most of its stern. Bert Edmonds got killed that way, making Torpedo 8's ninth fatality since Midway. The rest of the men on the destroyer helped in the saga of its homeward voyage. But that's another story, a remarkable one, but not a tale of torpedo bombers.

"Well, fellows," said Swede wryly to those who were left, "you are going to have something to tell your grandchildren after all."

He had been given tommy-guns, rifles, and hand grenades by the Marines and he distributed them to his men. Then he moved them up to the line where Major Mahoney, Captain Aaronson, and Lieutenant Wallace welcomed them hastily.

They were given secondary positions and Captain Aaronson helped Swede space his men and told him what to watch out for.

"It's very ABC around here," he said. "Just wait until you see a Jap. Then kill him."

That was three o'clock in the afternoon. There wasn't much to do until nightfall. With the night would come the malarial mosquitoes and the Japs.

After the foxholes had been dug to command fields of fire, Swede lay down and

wrote a letter to Missy, summing up with masterly brevity his paramount needs of the moment. "When I get home, please buy a brewery and some cows. I'm going to drink the brewery dry and eat the cows raw."

Fighting on land always seems a little unnecessary to aviators. But it's better for morale than not fighting at all, decided Swede. When the Jap planes came over low, as they did now, almost with impunity, all the fellows shot at them with their rifles and tommy-guns. They didn't seem to hit anything or do any damage, but they made a lot of noise and cheered themselves up.

Peterkin and Hammond and eighteen men were back on the field working on the three planes they hoped would fly again. They dug foxholes for themselves right alongside the planes, so that they wouldn't have to waste too much time going to and from cover during air raids, and they never dropped their monkey wrenches until the bombs actually started to fall.

The rest of the squadron helped hold the line until reinforcements could come. The jungle where they were was very thick. The light that managed to get through to it was green-colored, and the ground on which they lay was squashy and steamed with the rotted vegetation of centuries.

As long as there was daylight, the jungle was as quiet as we were willing to allow it to be. Our gunshots went through it like snores through a dormitory, leaving silence in their wake. But at night the whole jungle came alive, with Japs, animals, and gunfire. It was impossible to see anything, and the sounds you heard came from all around you. The tickle on the nerves was constant.

Swede lay peering at the darkness with a boy named Rich, who came from St. Louis and had worked in a brewery before becoming a bombardier-gunner. Rich talked steadily throughout the night. He was nervous and the words just wouldn't stop coming out of him. He talked about St. Louis winning the pennant. Then he talked about the brewery. When he talked beer, he was really good. Swede could feel it pouring into his ears and seething and frothing in his mind.

But maybe that was because a cold drink was so hard to come by on Guadalcanal. There was only one place to get it actually. That was at the ice plant the Japs had left behind when they ran away. There was a big sign on the ice plant: "TOJO ICE CO. NOW UNDER NEW MANAGEMENT, W. J. GENUNG, PROP."

And the fellows there would dip your canteen into ice water and freeze it up for you fine any time you came there. But the difficulty was finding time to go there, and then, in that simmer of sun and on the weathered-up days, that clutter of moist heat warmed up your canteen very fast. Rich had a monotonous voice, and the only times he stopped using it were when the gunfire was extra loud.

Then he'd listen nervously to the gunfire, and, after it stopped, he'd start to talk again, nervously. When he finished with the brewery, he talked about home, and what his girl said to him and what he said to her.

They seemed to have had some very remarkable conversations, in all of which the girl got the best of it, and managed to prove herself the best, sweetest, kindest, altogether most wonderful girl that ever had breathed the air of St. Louis. When he had nothing else to talk about, he talked about the dresses she wore. She had quite a wardrobe and Swede learned about every stitch in it.

Swede dozed steadily.

The rumbling sound of the guns seemed lulling after a while and he dreamed fitfully of mailing his letter to Missy and tried, in his dream, to dream of Missy reading the letter, but instead dreamed that he couldn't dream it.

Whenever the Japs fired their eighty-one-millimeter mortars, Swede woke up. They made a noise that made you think your head had been knocked off.

12 - Back into the Air

Major Mahoney, who was living his last days while the plane in which he was to die was being repaired by Hammond, and Captain Aaronson, who was to distinguish himself shortly when wounded by getting off his stretcher to knock out some Japs bothering friends of his, were the very best of hosts.

The fellows in Torpedo 8 remember that distinctly. Then they remember that, on the morning after their first night of jungle fighting, they saw nine or ten Japs sleeping about two hundred yards away. One fellow got excited and before warning any of the others threw a grenade without removing the pin. So, all that Torpedo 8 saw of those Japs were their backsides scuttling into jungle thickets.

But mostly their land fighting is recalled by them as a blur of shooting and throwing grenades, in the direction of sounds, though they couldn't see who was making them. During the day, they'd go back to the airfield to see how the maintenance men were making out with the planes and at night they'd go up and drowse through the gunfire on the line.

The business of land fighting seemed, at first, to fascinate only R. T. Williams, a young ordnance man who fancied himself with a rifle. He said there was one thing better about land fighting than air fighting. You could count your dead on the ground.

"You can if you ever see them," the rest complained.

And Williams said he'd keep at it until he could see them and count them. Finally, he did just that. A sniper made the mistake of showing himself and Williams killed him and counted him.

So, then Swede agreed that it could be done, and he and Hammond, Price, and Williams got into a jeep and motored up to the lines. There had been costly fighting and the sick-sweet smell of the unburied dead was thick among the trees.

The fellows smoked cigarettes constantly to keep their stomachs from turning. The jeep poked and snorted noisily through the jungle path.

Then snipers cut loose on both sides of them and all four of the men were out of the jeep and behind trees in a single motion; another one of the numerous reflex actions known by the generic term of "Guadalcanal twitch."

They stayed behind the trees for a few moments. The snipers were silent. The motor of the jeep idled gently in the middle of the path. That was the only sound to be heard.

"Well," said Swede, "we wanted it. Let's get it."

"Okay," said Hammond. "If you need a monkey wrench, call on me."

Williams, who was now the expert land fighter of the squadron, said that all they had to do was find out where the Japs were, and shoot them. And how was that to be done? Well, that was simple, too. One man would stick himself out and let himself be shot at. The others would watch for where the shots were coming from. The fellows thought that over awhile. Then Swede stuck himself out cautiously from behind the tree. He heard a shot and jumped simultaneously. The shot hit a telephone wire and a spark leaped a foot into the air.

"What do we do now?" asked Hammond. Nobody had seen where the shot had come from. They all had been closing their eyes, praying for Swede.

The business of drawing fire went on for about half an hour.

All four of the men prowled around cautiously, peering until there were green flickers in their eyes. The Japs had automatic rifles with long brass clips of bullets, a kind of very light machine gun.

Sometimes they'd let a whole clip go and sometimes be content with one bullet, but they never fired when anybody was where he could see them. They were canny, cautious, hardened veterans of jungle fighting, and this fact began sinking into Swede, as he realized how little he himself knew of the art.

"Well," he said at last, "those guys are using their brains. Let's use ours. Let's get organized around here."

They were all willing.

"Let's look for trees that look phony, that are built up too thick, and let's just figure that's camouflage hiding a gangster in it." (Gangster was the term the Navy used for Jap planes.)

Then Swede went to work with his field glasses. He studied the treetops carefully and finally found three that didn't look right to him. They all went to work on those, firing systematically to cover every square inch of surface.

Not a sound came out of those treetops, not a groan or scream or murmur. But what was more important, not a rifle shot came out either, and, after a while, the four men walked erect to their jeep and drove back home.

It was a relief for Torpedo 8 to get back into the air. The first plane to come off the Peterkin-Hammond reassembly line looked like a homemade bird. It had wings from one plane, an engine from another, and patches of tail from a third.

Hammond and Ike Hallam both agreed it wouldn't fly. One of the wings seemed out of line to them. But Swede tried it, anyway. The battle situation was still such that anything at all was better than nothing. Swede used up the whole runway for his take-off, and then, when he finally wabbled into the air, he found that one wing dragged a little. But at least he was flying.

When he got back, Hammond shook his head with mock regret. "I bet you wouldn't make it," he said, "and now I'm out money."

The Japs hammered Henderson Field relentlessly from the air by day and by night, and, day and night, they hammered at it from the ground. Swede went after the guns again the next day. The problem was difficult because of the coral caves into which the guns could be withdrawn, and because there was no chance for a sneak attack. The Japs could see him take off.

But Swede swung wide and moseyed that way until he got into a cloud. Then he ducked behind the mountains and came around in back of the Japs. He caught them. He saw the muzzle of a howitzer and dove at it and dropped a bomb.

The bomb missed, but Wendt and Sparks had been working their machine guns and nobody tried to move the howitzer after that. Swede made seven passes at that gun, each time dropping one bomb. His plane was hard to fly. Being wing-heavy, he had to fight it all the time to keep it level. But finally, he was satisfied that he had got the gun, and went back to the field.

Swede made another attack on a gun position and, in the evening, Hansen took over the plane. It was a clear beautifully moonlit night, and it was believed the guns would show up in it. Or, if they didn't, their muzzle flashes would. His first trip out, Hansen hit some Jap ammunition and lit a fire that everybody on Henderson Field watched with interest. Then Major Mahoney, who had come down the line for a visit, said it took a land fighter to spot gun positions and offered to go along and help.

Francis and MacNamara made room for him and Hansen took off again.

They went along pretty high in plain sight of the field. Then suddenly they could be seen diving for an attack, disappearing into the darkness behind a hill. Everybody waited for the plane to reappear again. There were sharp sounds of firing and the rumbling crump of falling bombs.

Then silence. And still the plane did not come back up over the hill. Then there could be seen a flicker of fire, like a flaming onion, whisking briefly along the trees and disappearing into the sea. The Japs had set Hansen's plane on fire and Hansen, too low for anybody to bail out, had run for water and crashed his plane into the water before it had time to explode.

Hansen, Francis, and MacNamara got out of the drowning plane practically unhurt, but Major Mahoney went all the way down to the bottom with it.

The battle raged steadily from October 22 to 27, but our forces would not flatten, so the Japs broke off the action and set the Tokyo Express rumbling to bring in reinforcements. Torpedo 8 and the Tokyo Express got going in different directions on the same track at the same time: five-twenty-five, November 7, a Saturday night.

There were eleven enemy ships in the party. There were twenty enemy planes there, too, and Torpedo 8 consisted of three planes—Divine's, Evarts', and Larsen's—although Torpedo 8 was not all there was of American contribution to the festivities. Major Joe Sailer (now listed as missing in action) brought Marine dive-bombers along and there were fighters, too—just how many is not clear, since several got lost in the weather, and never did show up in time.

It was raining when the fellows got the word to go and started off down the hill to their planes. A machinist's mate was fixing himself up a foxhole fit to stay the night in. He was spreading a poncho over the mud on the bottom and some Navy foul-weather issue clothing over the top.

"You'll be better off in your tent," said Swede.

"No, sir," the boy insisted, "not with Japs coming."

"Every time you turn over in your sleep, you're going to put your face in the mud."

"That's right, I guess," the boy said and went on with his housekeeping arrangements. "But sleeping here I don't have to worry if the first shell will get me. I can't sleep if I have to think all the time, will I get up fast enough to run away from the first shell."

"Hell," said Swede, "there aren't going to be any Jap shells tonight."

Everybody seemed to think different. Everybody was fixing up foxholes against the rain coming down, and the fellows going to their planes saw them doing it and thought somberly, "They seem to have plenty of confidence in us."

The weather was bad all the way up there, but then our side had luck and the skies where the ships were showed light and were feathered over with clouds. But the Jap planes had stationed themselves along the clouds, and were patrolling the clouds like angry dogs patrolling a gate. Planes seem to go slow when coming in for attack.

They are hurtling along, of course, the fastest they know how, but they look slow, sedate, and unmoved at the beginning, maybe because the mind is working so feverish fast that anything that is going to happen seems to be taking a long time to do so.

And our planes went slow, sedate, and unmoved-seeming toward the clouds and the Japs, while the twenty waiting Japs untangled themselves from their clouds and came prancing to the meeting. The torpedo planes were lower than the others and to the rear. They were concentrating on the ships. The ships were spread out in the shape of a fan. There was a heavy cruiser among them and that was the ship the torpedo planes decided to get.

Swede led the attack. As against the other cruiser on that August twilight which now seemed so many lifetimes past, there was a cloud hanging behind the target out of which the planes could dart and be right on top of where they were going. Swede headed for this cloud. Then suddenly he saw Joe Sailer cut athwart him, Zeros on his tail, and signaled him to join up.

But Joe wanted to solve his own problems without interfering with the torpedo attack, and refused to join up, and instead zoomed upward, his rear gunner flailing out and all around him.

The brawl going on upstairs—as the aviators call it—tumbled down the stairs now and was all around the torpedo planes. An air fight, Roy Simpler once said, after coming out of one of them, is like a dozen bees in a glass jar.

Swede and Evarts and Divine went steadily through all the bees and Divine had bad luck. Two Zeros jumped him.

Divine had R. L. Shively as his tunnel gunner and C. E. Monroe as his turret gunner. Monroe got one of the Zeros and saw it hit the water and was looking around for the other Zero when Shively screamed, through the radio, "Oh, the dirty goddam bastards," and the plane dived into the cloud. "I'm hit," said Shively.

They all kept going. That was the theory they had; whatever happened, to keep going until stopped. But they were not in formation anymore. The dogfight had split them up so that each man was on his own.

This Jap cruiser was craftier than the other had been. As soon as the torpedo planes went into the cloud, it swung over toward it, too, and was waiting under it when Swede came out. It threw up a whole fanfare of stuff to greet him, and Swede saw that he was too close to drop and made a big U-turn back into the cloud. It was like a thing by Charlie Chaplin with Chaplin, on the run, getting into the wrong place, jittering around on one foot and running out again as fast as he had come.

But it didn't seem funny to Swede at the time and it didn't seem funny to the Japs, who must have known how ordinarily fatal it is to give a plane two chances to get a bead on you. There was an unworldly stillness in the cloud. It seemed as remote from the battle on its edge as heaven is from hell.

The stillness came right up to the plane and the propeller thrashed at it and ground it up and seemed to mince the cloud into a thin spume, until finally there was the cruiser, plain before Swede, spitting death at him. Thirty million dollars' worth of cruiser. Thirty million dollars' worth of death. Swede pressed the trigger and swung and ducked low.

The anti-aircraft fire climbed up into the air where it had expected him to be and then dropped low as he booted himself high into the sky and got himself swallowed up in the unearthly stillness of the cloud.

"Shively," he heard, in Monroe's voice, "where are you hit?"

"It could be worse." That was Shively's voice, tired-sounding, but calm.

"Yes, but where?"

"I got it in the arms and legs."

"Which arms? Which legs?"

"There's slugs in both arms and in both legs."

"Gosh, and you said . . . What could be worse than that?"

"Well, I might have been born a freak with three legs."

Then Swede was out of the clouds. So, Divine's plane was all right. He looked down. The cruiser was crawling like an animal with a broken back.

There was a destroyer with flames coming out of its starboard side and fantail, and far off was a torpedo plane, hugging the water low with tracer shells racing after him and dropping short of him.

That was Evarts, getting away safely. The water was pocked with the corpses of burning planes. From each plane rose a column of black smoke, going up straight into the air like some gaseous breath. There were thirteen dead planes there, one of them ours.

Then, just as Swede was reaching the horizon, he heard two explosions, tremendous enough to reach over all that water and carry above the noise of his own plane and into his sealed cabin. He looked back, startled. The explosions, one on top of the other, had come from the cruiser. The cruiser was enshrouded in smoke.

When Intelligence got the score figured out, they gave the dive-bombers hits on one destroyer and counted all three of Torpedo 8's wallops as having landed—two on the cruiser and the third on a second destroyer.

It looked very good to Swede and he went down the hill to the foxhole where the machinist's mate was sleeping. "Wake up," he said jubilantly, "and go to sleep in your tent."

13 - FAREWELL TO GUADALCANAL

AT THE END, THE JAPS GAVE Torpedo 8 a visual demonstration of a torpedo bombing attack that made the fellows very thoughtful. Some of the men were down on the beach November 12 to see if any of their friends had come in on the four transports there, when the Japs threw a force of thirty-three planes at the vessels, most of them torpedo bombers.

From twenty to twenty-five of the planes were two-motored Mitsubishis of the torpedo-carrying variety. This made a force as large as the Japs had thrown at the bulk of our Pacific Fleet at Pearl Harbor, December 7, and a force at least three times bigger than any Torpedo 8 had used in any single attack since Midway.

It was a gray day. There was an overcast, and mottling it were huge accumulations of mouse-gray clouds that made perfect ambush territory for pilots. But only one of the Jap planes lived through the attack and that one was a Zero, not a torpedo bomber.

The twenty to twenty-five Jap torpedo planes managed to live long enough to launch four torpedoes. But these torpedoes hit only the water. They did not hit any ships. The only damage done was when a Jap torpedo plane exploded so close to the cruiser San Francisco that its debris killed eighteen men and burned several others. The attack was a complete catastrophe for the attackers. The Zeros came over above twenty-five thousand feet, trying to pull our fighters up there and keep them off the torpedo planes. They succeeded, but not for long enough. They got themselves shot down too fast.

And then our fighters dived or the torpedo planes. Our fighters were intent and eager. They plummeted for their prey at terminal velocity.

The case of Captain Joe Foss, a Marine ace from South Dakota, who raised his "kill" to twenty-one planes in the course of the day, was typical. He shot down a Zero at twenty-nine thousand feet, dove down to less than three thousand feet, and ended the careers of two torpedo bombers. He dove so rapidly as to defy the laws of physics. The wind stirred by his dive stripped the glass from his shelter hatch and tore off all the rubber pads on the outside of his plane, but he was working on his third victim before his first had completed its fall into the sea.

As soon as the alert came, the four transports and their escort pulled away from the beach. They were traveling in column abreast about six or seven miles from the beach when the Jap bombers came out of a cloud about four miles down the beach from where the ships had been.

They had used a cloud, but they had used the wrong one, and from where they were, they had a run of almost two minutes to their targets, and, in addition, had to run a gantlet of fire from the beach. This was not yet an inevitably fatal mistake. It was a mistake, but it could have been redeemed. Torpedo 8 had made runs that were longer than that and had submitted themselves to even more concentrated gantlets of fire when necessary.

The thing about this was, it wasn't necessary. There were clouds near the ships that could have provided the Japs cover until the last seconds before the assault. And when the Japs came out of their cloud and saw that their leader had led them into a serious mistake to which they all might succumb, they all, in one way or another, gave way to despair.

One actually circled so that his plane looked like a man wringing his hands and kept circling until a Grumman fighter blew him up. Others scattered out of formation and became easy prey. Still others just gave up jinking and flew to their deaths as if frozen in fear. It took less than ten minutes for the sound of all firing to stop and a great unnatural quiet to settle upon the gray-colored arena.

At the end of that time all the Japs were dead except three; one in the Zero on his way home with such thoughts as are inevitable in a man who lives alone where all his friends have died, and two others who had survived the crash of their plane and had allowed themselves to be taken prisoner, after killing their officer when he tried to prevent them.

Going back to the airfield, Hammond shook his head.

"Gosh," he said, "is that the way you guys make your living?"

"Hell, no," replied Swede. "If that's what we did for a living, we wouldn't live any longer than they did."

"Transports is hitting sitting ducks for us," said Evarts. "Those fellows just threw themselves away."

"They lost their nerve, that's all," said Swede. "If they had just kept on doing what they had to do, they'd have had six or eight seconds start on those fighter planes. All those fighter planes were sucked out of the play at the beginning by the Zeros."

"They came back into it fast," pointed out Hammond.

"Yes, but in a good torpedo attack fast is never fast enough. If you're a fighter trying to break up an attack, you got to be there when you have to be. Any time you spend getting there is too much time."

"That is, when you fellows are doing the job."

"Yes, us, not those dead Japs out there."

The torpedo action was one phase—a minor phase, as it turned out—of the fifth in the Battles for the Solomons, an engagement now known officially to the Navy as the Battle for Guadalcanal.

This battle was fought out primarily by warships in two major night encounters and one minor one.

In the daytime, airplanes took over and killed off the ships crippled by warships the night before, and killed the transports the warships had been supposed to guard. Navy, Marines, and Army fought the battle for us, and their team-play was as flawless as that which goes on in a watch—one part taking over from the other, one part moved into action by the other, and all parts together tolling off the last minutes for untold thousands of Japs. The Japs clung relentlessly to their tactics of trying to knock out our air-power by bombardments from the sea before moving in with their troop-laden transports.

They came to do it with battleships on the night of November 12-13 and again on the night of November 13-14 and a third time on the night of November 14-15. Each time our Navy met them and smashed them in a disaster which is as yet unprecedented in the history of great navies. The Japs lost twenty-eight ships—sixteen warships (among them two battleships), eight transports, and four cargo vessels—three ships more than the combined loss of the British and Germans at Jutland.

During those inexorable November days and nights, spotted bloodily by hours and half-hours in which thousands of men died, Henderson Field buzzed and swarmed with activity. Reinforcements came from everywhere and for

everybody, for the Army, the Navy, and the Marines, and even the inveterate crap-shooters on Guadalcanal got reinforcements.

The land forces had very little to do in the battle, and the influx of fresh money and new talent, combined with the unusual amount of leisure time available, stirred the sporting fraternity as they had not been stirred since August 7.

In the middle of the battle, Torpedo 8 got the news that its long-awaited relieving squadron was on the way. But, in the meantime, it still had some loose ends to pick up. A Jap battleship, crippled by cruisers the night before, lay smoking and floundering within sight of Henderson Field. That was November 13, a Friday, and a little after noon, Larsen, Evarts, Divine, and Engel, joined by two Marine planes, went up to put some torpedoes into it. The problem was comparatively simple.

Although five destroyers were guarding the battleship, there was good cloud cover near it. But Swede was not the man to take chances, not on a Friday the Thirteenth on what was likely to be his last attack mission before going home. So, before taking off, he synchronized watches with Major Joe Sailer who was leading dive-bombers against the ship.

The writer saw Swede go off on that attack and saw him come back a few moments later and then took his picture. Swede was carrying his gear when he left. His hair was freshly brushed, and he was freshly shaved, as a man should

be when he starts off to work. He was smiling, too.

The news of the battle was all good. The Navy reinforcements that had come in included many old friends of his, and there was so much gossip back and forth about what So-and-So was doing and what So-and-So had told the Commander, that a man had no time to think of what lay ahead of him when he got into the air.

Swede was smiling, too, when he got back. His hair was only a little rumpled where his helmet had sat, and he seemed eager as he came up the hill to the ready tent. He said he was eager to hear more gossip. The chances are against that being exactly true. Young men are resilient and recover very rapidly from the shocks of battle.

But they are not that resilient. They do not have rubber bands for hearts. And for a second there, when he was very close to the battleship, it had looked to Swede as if he were not going to live to get his leave. The approach to the target had been planned perfectly. When Swede got into the cloud he had picked as his springboard for the assault, he heard Joe talking to him.

"Mark one," said Joe, and Swede, loitering, held his wrist watch up before him. "Mark two, Mark three, Mark four, and . . . and . . . and go!"

Then Joe went and Swede went, their formation following on behind.

A Jap gunner got a bead on Swede. Swede knew it, but he didn't panic. He remembered the Jap torpedo planes of the day before. He saw Joe's plane diving, and thought, this was split-hair stuff, this was walking tightrope on a split hair, and kept throwing his plane around, jinking and corkscrewing. That Jap gunner wouldn't get off him.

He followed Swede's every move, and Swede knew that when he leveled off to drop his torpedo and gave the man a steady target, it was going to be bad business. But what had to be had to be, and Swede leveled off at the time he had planned to, and at the time he had figured on with Joe, and dropped his torpedo. Then nothing happened.

Swede got away all right, as did all the others, and Joe and the dive-bombers, too. An officer, flying over the scene as an observer in a scout plane, reported that Joe's thousand-pound bomb had hit the battleship amidships a few seconds before Swede's torpedo had hit it there. The torpedo had gone in right under where the bomb had hit. Which meant that Joe's bomb had wiped out the gunner who had been about to get Swede. When Joe and Swede heard this, they said, "Well, why not? We've got very good watches."

The last day Torpedo 8 spent on Guadalcanal was November 15, a fine, gentle, very sunny, windy Sunday like a late spring day back home. And Swede got up a picnic for his men. He took all of Torpedo 8 that was left on the island—Peterkin, Engel, Hammond, Hallam, Pop Lawrence, Rich, Bartlett, King, and Liccioni—for a sight-seeing sky tour of the battlefields. Two planes from Scoofer Coffin's squadron (Lieutenant Albert D. Coffin of Indianapolis) went along to add to the number of torpedoes carried.

There were seventy miles of wreckage to be seen along the groove—debris of ships, corpses, oil streaks, life rafts, and an occasional burning hulk. Then they found a merchant ship, swinging idly as if abandoned, and Swede said it was worth one torpedo and he'd deliver it.

But he was in too relaxed a mood, or something. Anyway, he made what he describes as a bum-run and missed everything, but the horse laughs directed at him. After that, Engel said to watch him, and Swede went over the ship and hovered there while Engel made his run.

"Come on," yelled Lawrence, "knock that bum out of the park!"

"In the la bonza!" shouted Liccioni. "Right in the old bread-basket."

Engel's torpedo hit the ship in the stern.

Looking down through the clear sunlit air, Swede could see the explosion come out simultaneously on both sides of the ship, ripping its stomach off all around, and saw unexpectedly a man take a running dive over the side into the water. Some Jap had been there, "playing possum" while the planes frolicked about him, and maybe, too, there were other Japs there playing possum.

The planes stayed on the scene until the ship sank. It went down in a very few minutes and the fellows watched fascinated. Of all the ships they had hit, and others had seen sink, this was the only one they themselves had actually seen go down. Some more torpedoes were put into the hulks of burning transports, just in case. Then yet another freighter was found swinging idly, apparently abandoned, and Ensign Wells, one of Scoofer's boys, who had the last torpedo left, went down to take care of it.

Wells made a very good run. He gave a real schoolbook performance, and in the clear water the yellow warhead of his torpedo could be seen racing along like the business. The torpedo hit the side of the ship, bounced, swerved up the length of the ship, and then sank. It had been a dud.

"Well," said Hammond regretfully, "I see where us guys who stay on the ground are some use to you."

"You are when you are," said Larsen, "and when you're not—holy murder!"

When the fellows got back to Honolulu along about Thanksgiving, some radio company put them on the air for a broadcast and sent messages to the home folks advising them to listen in.

Missy got one of the telegrams and grouped her children and herself around the radio. She sat for a half-hour waiting. The children became restless and Swiss wandered off and Melissa crawled off. Then she heard Swede's voice. She sat frozen.

She had not really believed he would ever be safe again. Swiss came running, wide-eyed, back into the room.

The telephone started to ring. It was friends who had tuned in on the program by accident and were afraid Missy did not know about it. Missy did not see Melissa standing there. She did not hear the telephone. She did not even hear the words Swede was saying. She just heard his voice.

THE END

Made in the USA
Coppell, TX
07 January 2020